W9-AXR-852

Scientists and the State

Scientists and the State

Domestic Structures and the International Context

Etel Solingen, Editor

Ann Arbor

THE UNIVERSITY OF MICHIGAN PRESS

Published in the United States of America by
The University of Michigan Press
Manufactured in the United States of America

1997 1996 1995 1994 4 3 2 1

A CIP catalogue record for this book is available from the British Library.

Library of Congress Cataloging-in-Publication Data

Scientists and the state : domestic structures and the international context / Etel Solingen, editor.
p. cm.
Includes bibliographical references and index.
ISBN 0-472-10486-1 (alk. paper)
1. Science and state. 2. Scientists—Political activity.
3. Science and state—Case studies. 4. Scientists—Political activity—Case studies. I. Solingen, Etel, 1952– .
Q125.S4365 1994
338.9′26—dc20 93-44407
CIP

To my mother, and the memory of my father,
the true source of my scholarship.

Acknowledgments

Many debts have been incurred in the process of organizing this book. The conception and definition of the project would not have been possible without the initial support of the University of California's Institute on Global Conflict and Cooperation, through a Sloan Foundation Postdoctoral grant. This stage was carried out under the sponsorship of UCLA's Center for International and Strategic Affairs. UCLA's International Studies and Overseas Program selected the project for its First Postdoctoral Fellowship program and hosted a workshop in January of 1990. The workshop was largely funded by a grant from the Columbia Foundation while most administrative tasks fell primarily on the shoulders of UCLA's Latin American Center. UCLA's Center for Russian and Eastern European Studies hosted the contributor of the Soviet case study. Behind this long list of institutions are the real sponsors of the project: I owe special thanks to Herbert York and the I.G.C.C. staff, John Hawkins, Michael Intriligator, Arthur Singer, Johannes Wilbert, Norris Hundley, John Hatch, and Norma Farquhar.

At various points in the process the careful comments of Larry Badash, Regis Cabral, Karl Hufbauer, Sanford Lakoff, Robert Merton, Judith Reppy, Norman Vig, and Richard Worthington helped improve the conceptual framework that guided the country studies. A revised version of that framework entitled "Between Markets and the State: Scientists in Comparative Perspective" appeared in *Comparative Politics* 26, 1 (October 1993). The case studies benefitted greatly from critiques by workshop discussants Ken Conca, Roman Kolkowicz, Fred Notehelfer, T. V. Paul, and Norton Wise. Felicia Yu coordinated the production of the manuscript in her impeccably professional style. Claudia Arias, Richard Siao, and Maria Gomez were the most effective research assistants for whom anybody can hope. Vicky Ronaldson, Ziggy Bates, and Cheryl Larsson are not only masterful in word processing, but also in telemetric decoding. Finally, my husband, Simon, and children, Aaron and Gabrielle, are not only a wonderful source of moral support but also helped keep my interest in science and politics in proper perspective.

Contents

Domestic Structures and the International Context: Toward Models of State-Scientists Interaction

Etel Solingen

This book examines patterns of interaction between modern states—in their varying forms—and scientific communities.[1] The book originated in an attempt to study broad patterns of political relations involving scientific communities too often ignored by science-policy analyses. That initial conceptualization was rooted in a social-scientific, rather than a history of science, tradition and sought a more systematic understanding of historical processes shaping state-scientists relations.[2]

The more formal aspects of the original conceptual framework were later relaxed to allow a more interdisciplinary perspective. Yet the project retained a heavily comparative analytical flavor, stemming from the puzzling observation that state-scientists relations vary across political-economic systems despite a universal appreciation for the instrumental role scientists play in the attainment of state objectives. The case studies in this volume generally explore the extent to which the internal structures of political-economic systems, on the one hand, and the nature of their involvement in the international context (global, regional), on the other, may influence the development of different models of interaction between states and their scientific communities.

This effort differs from earlier studies about the relationship between science and politics under various political systems in a number of ways. It defines *political* relations as the main variables to be explained, explores a

---

1. I use *scientists* and *scientific community* interchangeably here, without implying that scientists are bound by spiritual relations prevalent in a Gemeinschaft (Haberer 1969), but merely by common professional norms and ties. As argued, "the republic of savants is neither one nor indivisible and no more constitutes a united family than any other professional group" (Salomon 1973, 167).

2. Solingen 1989. This conceptual framework guided the drafting of preliminary papers—presented at a UCLA workshop in January 1990—which later became the chapters in this volume. A revised version of the original framework appeared in Solingen 1993a.

structural understanding of those relations, provides a more systematic *comparative* framework than attempted hitherto, and incorporates the *international political and economic sources* affecting state-scientists relations. It does not, therefore, concern itself with the comparative social-political context of scientific creativity.[3] Comparative and historical macrosociology of science—a respected tradition rooted in Merton's structural-functional approach—has been mostly interested in cultural differences and the internal norms and output of scientific activity.[4] Its interest in political determinants was relevant mostly to the task of identifying a compatibility between certain political and economic systems, and the scientific enterprise itself. Rarely were the *political* relations between scientists and the state, rather than the content of science, the main dependent variable in this tradition.

The political science literature that flourished in the 1960s aimed at understanding the political role of scientists by concentrating on the concept of scientific advice.[5] The generic aspects of the interaction between states and scientists—which are broader than the realm of scientific advice and encompass a more collective, structural understanding of scientific communities—are significantly absent in these efforts. Yet the unprecedented growth of scientific communities—an important segment of what Reich (1991) labels "symbolic analysts"—compels such collective perspective; there are more scientists alive today than there have ever been in all previous generations together.[6] Moreover, that literature was circumscribed, to a considerable extent, to the U.S., British, and Soviet experiences. Thus, in his seminal volume *Science and Politics,* Salomon (1973) argued that empirical studies of the political relationship between scientists and the state were rare, limited in scope, and restricted in their themes. Haberer's (1969) preface to his *Politics and the Community of Science* reflected on the paucity in truly comparative analysis: "The task of creating the theoretical foundations for the comparative study of science and politics remains to be done" (p. iii). In a comprehensive

---

3. Neither do the chapters focus on philosophical questions about the role of science under capitalist or Marxist modes of production posited by critical social theory, including the work of Marcuse 1964, Habermas 1971, and Horkheimer and Adorno 1973.

4. See Ben-David 1978. Representative studies in this tradition include Barber 1952, Kornhauser 1962, Kuhn 1962, deSolla Price 1963, Storer 1966, Merton 1967, Shils 1962, Merton 1973, and Ben-David 1971.

5. See Kramish 1959, Snow 1961, Gilpin 1962, Gilpin and Wright 1964, Jacobson and Stein 1966, Lakoff 1966, Holloway 1970, Gowing 1974, and Kevles 1978, among others, followed by a prolific output since the 1960s, including Golden 1988 and, most recently, Jasanoff 1990.

6. Landes 1969, 517. Reich (1991) places scientists at the forefront of this economic category involved in problem-solving, problem-identifying, and problem-brokering activities, which can result in services traded worldwide.

review of the literature on scientists, technologists, and political power, Lakoff (1977) suggested that the field would be richer, and the opportunities for comparative study much improved, if more of the investigations were made with respect to other political systems.

Finally, studies in this tradition were deeply rooted in the understanding of domestic—mostly political—institutions and processes. Rarely were the broad international political and economic conditions that may shape the relationship between scientists and the state incorporated systematically into the analysis of contextual influences. A pioneering effort in that direction—Gilpin 1962, 1968 and Skolnikoff 1967—unleashed increased interest in the impact of nation-states' scientific and technological capabilities on foreign policy and international affairs.[7] We are more concerned here, however, with the impact of changes in the external context on the domestic political economy of science. In international relations theory, the first set of causal inferences (domestic to external) lies within what Waltz (1954) labeled "second image" or "inside-out" effects. The second set explores "outside-in" (external to domestic) influences and is known as "second image reversed" processes.[8]

Each of the chapters in this volume is broadly organized around the major theme, analyzing the impact of state structures and state involvement in the international political and economic systems on the domestic political economy of science. The authors provide a synthetic review of the resulting internal characteristics of the scientific community, and of historical patterns of state-scientists interaction in each country. Against this general background, most of the authors here place special emphasis on the case of physicists and discuss the core domestic and international issues that trigger the politicization of the scientific community—nuclear energy, arms control, human rights, and economic development.

A parsimonious theoretical framework, attempting to order a wide variety and number of observations while relying on a discrete number of theoretical propositions, is often overly constraining. The case studies here, therefore, although guided by a general common framework and loosely conceived as empirical plausibility probes, were not designed to test a comprehensive and fully developed body of propositions.[9] They do suggest, however, interesting variations across political-economic systems.

---

7. For a comprehensive review of this field see Schroeder-Gudehus 1977.

8. Gourevitch 1978. Beyond this focus on subnational phenomena, Ruggie (1975) explored international responses to technology; Haas (1975) examined the relationship between knowledge and the construction of international regimes; Haas, Williams, and Babai (1977) studied the world views of scientists working in international programs; and Nau (1972) analyzed interdependence and scientific cooperation in Western Europe.

9. See Eckstein 1975 on comparative case-study research.

The Domestic Structures of Scientific Activity

> It is the structure of political and economic organization which determines . . . the incremental rate of growth in knowledge and technology. (North 1981, 17)

This volume attempts to come to terms with the assumption that scientific organization, and derivative models of politicization of scientific communities, may reflect prevailing political and economic structures or state forms.[10] States have become nearly everywhere the main patrons of basic research—particularly where private benefits are low and public benefits are high. States, further, have exercised considerable control over detailed public allocations for science; defined topical research boundaries; steered private investment in science, to some extent or another; and regulated degrees of scientific interdependence with the outside world. States have relied on science and technology to secure their political, economic, and strategic viability.[11] The relationships between science and political power are, therefore, at heart, about the instrumental nature of modern science. The state's role in science grew also in response to the nature of scientific enterprise in the second half of the twentieth century and the short lead time between discovery and application (Gilpin 1968). The growth in scale and in the organizational complexity of science compelled states to plan the funding and training of scientists.

Defining the political priorities that are to guide scientific activities involves distributional choices. Scientists may be allowed to capture a concentrated share of benefits through employment, research and development (R&D) allocations, and various rewards. They may also be forced to absorb a disproportionate share of risks; travel restrictions in authoritarian contexts, for instance. The nature of the state is thus a useful point of departure in trying to understand the political nature of scientific communities. The organizational

---

10. *Politicization* here follows Ruggie's (1975) usage, pointing to the process by which phenomena are brought into the public domain and are, in turn, shaped by public considerations. The *state* is a contested concept; a working definition adopted among participants at the organizing workshop, which led to this volume, points to the bureaucratic, legal, productive, and coercive organizations through which policy makers operate.

11. Brooks (1980, 66) defines science as "conceptual knowledge involving mental models applicable in a large number of concrete situations." Technological knowledge, in his view can be scientific and abstract, but can also be concrete and empirical; it is often a mix, although new technologies tend to have a greater science component. Pure science and technology, in other words, lie at two extremes of a continuum of overlapping knowledge-seeking domains (Clark 1987; Paquet 1989). However, pure science does not always lead to technical advance while the latter can take place without science. In addition, problems related to technology can trigger new scientific knowledge. On the ideological foundations of basic science see Restivo 1974.

configurations of states are assumed to "encourage some kinds of group formation, and influence the political capacities, ideas, and demands of various sectors of society" (Skocpol 1985, 21). Nowhere is this more evident than in the case of scientists whose dependence on the state is almost unparalleled by other professions, even where liberal canons of economic organization prevail. Quite surprisingly, as Cozzens and Gieryn (1990) argue, studies of "science in society" have ignored the state as an analytical category.

We can derive a crude fourfold typology of twentieth-century states, as in table 1, from Lindblom's (1977, ix) contention that "aside from the difference between despotic and libertarian governments, the greatest distinction between one government and another is in the degree to which market replaces government or government replaces market."[12] Thus, we associate pluralist market-oriented systems, also known as liberal democracies or polyarchies, with free and self-regulating markets, private ownership of the means of production, rule by many through orderly and contested elections, and the exercise of individual freedoms.[13] Instead, noncompetitive centrally planned systems, allow neither market nor popular control over public decisions, canceling, to different degrees, free economic and political competition. They subordinate economics to politics and place means of production under public control. The historical versions of this ideal type are the former Soviet Union and the Eastern European countries recently dismantled by democratization and economic liberalization.

Developing countries have followed variants of these two main models. At times, they have combined waves of market-oriented economic policies with noncompetitive political systems. This was the case with a number of newly industrialized states, including South Korea, Taiwan, Brazil, and Argentina. At others, they have built more or less centralized economies around principles of pluralistic political participation, as in India and Israel. Earlier historical forms of market-oriented authoritarianism include Nazi Germany and Fascist Italy.

Following a methodology of "focused comparison" (George 1979) this volume includes various instances of pluralist market-oriented cases (United States, Japan, Federal Republic of Germany, and France); noncompetitive centrally planned cases (Soviet Union, People's Republic of China), noncompetitive market-oriented cases (Nazi Germany, Brazil between 1964 and 1985); and pluralist centralized cases (India and Israel). The political evolution of many of these states throughout this century compels a dynamic

12. This typology identifies "ideal types," which, in contrast to "concrete types," are classifications denoting "combinations of elements that seem to fit logically from the theorist's point of view" (Eckstein 1964, 20). The systems presented here are no more than analytical abstractions; in the real world they tend to share more than meets the eye.

13. Dahl and Lindblom 1953.

TABLE 1. State Forms

Economic Organization	Political Systems Pluralist	Noncompetitive
Market-oriented	United States	Brazil (1964–85)
	Japan	Nazi Germany
	FRG, France	
Centrally planned	India (1940s–1980s)	Soviet Union (1917–88)
	Israel (1950s–1970s)	PRC (1948–1970s)

Notes: FRG = Federal Republic of Germany, PRC = People's Republic of China.

historical analysis and provides a unique opportunity to combine longitudinal (within-country) with cross-sectional (cross-countries) comparisons. Thus, the advent of democracy in Brazil in 1985 brought its model more in line with that of other pluralist industrializing states, such as India and Israel. As Schwartzman notes, in this volume, political restrictions on scientists ceased to exist, although the economic crisis delayed the crystallization of a new relationship. Pfetsch's study of Germany, also in this volume, provides us with three historical variants. The Weimar Republic was subject to internal chaos and external constraints; scientific research survived largely thanks to industrial support. The Third Reich brought centralization, state control, and very high shares of military R&D. The postwar Federal Republic of Germany, a classical pluralist market-oriented case, reversed all three.[14] Each regime reflected a different pattern of state-scientists relations.

Variations in state forms help explain differences in the extent to which scientific activities reinforce or challenge a particular political context. Different states tend to rely on alternative mixes of exchange, authority, and persuasion in their relations with scientific communities.[15] Most systems seem to converge in their use of exchange as a basic mode of interaction, although they differ on the substance (or currency) of exchange. Some offer high degrees of scientific freedom in exchange for loyalty and subservience; others rely largely on material rewards. Most nurture a well-rewarded scientific elite, often associated with defense industries. Reward systems may be built on more or less equitable patterns of pay-scale, as in pluralist market-oriented contexts, or on the cooptation of elite and discrimination of rank and file, as in

14. Abstracting from the three models it can be argued that the higher the levels of economic interdependence (exports/GNP), the higher the levels of private sector funding for R&D. Exports as a percentage of GNP were low under the autarkic Nazi regime (6 percent). They grew from 15 percent during the Weimar Republic to 27 percent during the FRG period.

15. On exchange, authority, and persuasion as mechanisms of social control see Lindblom 1977.

the former Soviet Union.[16] The growing recognition of property rights—of research resulting in a "private good"—is a special kind of reward. Public recognition as a reward seems to cut across systems, although it does not always come about as a result of an exchange, and may derive from scientific achievements and sociocultural valuations of science.

There is a wider range of alternatives to political rewards available to scientists in pluralist market-oriented contexts than in noncompetitive centrally planned contexts. If the prestige afforded by certain governmental or academic position is not a paramount preference, scientists may opt for materially rewarding private enterprise activities. The ability to exit from the system of exchange prevailing in noncompetitive centrally planned systems is constrained by the absence of alternative (private-sector) exchange structures. Although the theoretical possibility of defecting under authoritarian rule is there, the costs in personal security and welfare are extremely high, acting to impose conformity. The use of market rewards applies to noncompetitive market-oriented contexts as well, although the role of coercive authority and the limits on scientific autonomy prevail as control mechanisms. Attempts to exit from authoritarianism in the People's Republic of China, Soviet Union, Nazi Germany, and Brazil and Argentina in the 1960s—scientists fleeing political persecution and intrusion and seeking asylum elsewhere—are at the heart of twentieth-century brain drains. The latter can also follow economic hardship, as with scientists from the developing world and from the recently transformed societies of Eastern Europe.

In pluralist systems, science plays a special role as an instrument of persuasion. It can be marshaled to defend contending political perspectives and reinforce competition in a truth-seeking society.[17] Here, the executive and legislative outputs affecting scientific activities are highly permeable to political, economic, and professional influences from below. Subtler ways of handling dissenting scientists may be used, such as excluding them from military contracts and/or withholding funding, both of which were practiced in the United States in response to opposition to the Vietnam War and the Strategic

---

16. Elite scientists were well remunerated and granted access to privileged shops, better housing, medical facilities, dachas, cars, and Moscow residential permits (Kneen 1984). Salaries and prestige were lower in industrial institutes than in theoretical institutes, and they doubled or tripled with jobs in the military sector, where scientific instruments were also the best (Rabkin 1988).

17. Science in pluralist systems is generally something to be discovered (or uncovered) and, as Smith suggests in this volume, often reflects partial and limited, rather than any universal, truth. In noncompetitive systems, instead, science is the correct and definitive point of departure for a new social order. On the compatibility between the values of liberal democracy and those underpinning scientific inquiry see Polanyi 1946 and Merton 1973. Ezrahi (1990) provides a superb analysis of the political functions of science in the liberal-democratic state.

Defense Initiative. In general, however, the attempt to transform the structure of science is based more on gentle persuasion, as Gilpin (1968, 248) argues with respect to France, than on the effective application of political or administrative power. In this volume, Baumgartner and Wilsford highlight how entire corps of scientists are coopted into the French state apparatus, following their common socialization in the *grandes écoles*.

Persuasion in the form of ideological indoctrination thrived in noncompetitive, particularly totalitarian, systems. There, the critical role of science was to provide a means of legitimating value claims, to the benefit of a social class or race.[18] When scientists resisted such requirements, authority quickly replaced persuasion as an instrument of control of scientists. Here, the party, bureaucracies, and professional associations served as hierarchical transmission belts of centrally defined priorities—mostly "public goods" oriented research. The Soviet Academy of Sciences became an agency of the state, where the Party controlled ideology, appointments, and policy directions. There were, however, variations among authoritarian systems. While the Stalinist and Maoist centrally planned models claimed individual as well as institutional control, the more market-oriented Nazi regime tolerated a measure of individual noncompliance while turning the Academy into a political instrument. Authoritarian regimes in the developing world threatened individuals without attempting, for the most part, to coopt scientific institutions.

In what ways do political and economic forms affect the loci of scientific activity? In noncompetitive centrally planned systems such as the former Soviet Union, most relevant research was traditionally conducted within the over 600 carefully monitored state institutes, not in universities. About 50 percent of Soviet scientists worked in the Academy and associated institutes, and the rest in ministerial institutes, state enterprises, and universities. State enterprises employed a very small percentage of R&D manpower. Among the consequences of this distribution was a lag in applying research results to industrial applications.[19] Although researchers at the ministerial institutes served industrial needs, their lack of proximity to plant requirements prevented effective contributions. Many of these constraints operate in developing countries that also tend to concentrate scientific research in state institutes—as in India, as Kapur documents in this volume.

The research structure of pluralist market-oriented systems concentrates R&D activities mostly in the private sector, and to a much lesser extent

18. On the planning of Soviet science see Graham 1967, Fortescue 1986, and Josephson in this volume.

19. See Graham 1977. Soviet plants were uninterested in the introduction of potentially disruptive new technologies, which truncated any attempt by theoretical and applied researchers to experiment with them. Measures to close this gap between industry and research included contracts between institutes and individual plants.

in government agencies and academic institutions. The proportions of total R&D financed by private industry are much higher in Japan (70 percent), and Germany (80 percent) than in the United States (50 percent).[20] Yet, as Baumgartner and Wilsford and Low suggest, even private research in France and Japan follows state objectives and strategies coordinated by powerful science-policy agencies. In fact, the share of scientific research conducted in public institutes places the French model closer to the old Soviet model than to any other market-oriented model. In industrializing countries like Brazil and India, investment resources available to the state, such as R&D allocations, are clearly more limited than is the case for advanced economies. Private sector investments in R&D are also generally meager; they are higher in Israel, with about 22 percent of the total, and employing 25 percent of R&D manpower. Most scientists in the developing world are employed by poorly funded academic institutions or by the state—a fact Kapur traces to weak market forces unable to create demand for scientific activities. Strengthening the weak links between private industry and research institutions has been a primary target of science and technology policy in countries like Brazil and India. The statist orientation in developing countries is regarded as a means to overcome underdevelopment and dependence, and to increase their power internationally.

The proportion of experimental scientists in industrialized countries (either pluralist market-oriented or noncompetitive centrally planned) is greater than in industrializing ones. Theoretical work is prevalent in the latter because there is no significant demand for scientists' output, either by state or by industry, and because experimental science is too costly for most of these countries.[21] As to the appropriate mix of pure and applied research, historical studies seem to suggest that noncompetitive systems (either market-oriented or centrally planned, developed or industrializing) tend to be particularly resistant to pure research. However, even the Nazi regime with its emphasis on applied research and technology was not completely prepared to repudiate purely intellectual pursuits, as Pfetsch makes clear.[22] Also in this volume,

20. "R&D, Invention & Competitiveness," *OECD S&T Indicators* (Paris), no. 2 (1986). Only 5 percent of government R&D funds go to industry in Japan; the rest to government labs and universities. Quasi-governmental public corporations in Japan (atomic research, nuclear fuel, and others) are jointly capitalized by industry and government and can enter contracts as private corporations (Long and Wright 1975).

21. Their most recognized scientists, particularly physicists from countries like China and Pakistan, are, for the most part, theorists, as Frieman (in this volume) points out.

22. On the attitude of totalitarian regimes toward pure and applied science see also Merton 1967, Haberer 1969, Hirsch 1974, and Beyerchen 1977. Official attitudes were clearly ambivalent and unstable; even concerns with national security could not offset Hitler's willingness to sacrifice scientific leadership in basic science by expelling Jewish scientists. Yet, the Kaiser-Wilhelm Gesellschaft, while purged, continued to be recognized as a critical instrument to

Josephson documents similar tensions throughout Soviet history. The ambivalence toward basic science—which accounted for 2.4 percent of all R&D expenditures in 1973—in the People's Republic of China was not a product of the Cultural Revolution, but preceded it.[23] Behind the ideological rhetoric about scientists' duty to immerse themselves among the masses, the Cultural Revolution was an extreme instance of the belief in the need to integrate science with social needs. Most recently, considerations of economic competitiveness have reinforced an emphasis on applied science, particularly because the Chinese leadership perceived Japan's success largely in those terms. Low's chapter in this volume analyzes the domestic and international sources of Japan's early postwar emphasis on applied science.

The International Context

Defining the role of science, particularly in the second half of this century, has become intertwined with processes beyond national boundaries. The rationale for exploring the influence of external circumstances over what amounts to an internal political relationship is rooted in the increasing inability of modern states to untangle those two domains of activity. Scientific activities have not only not escaped the secular process of global interdependence; they have been instrumental in strengthening it.[24] The international context can exert a powerful influence on the domestic conditions of scientific activity in two general ways.

First, the degree of external conflict leads generally to higher levels of investment in military-related scientific research, higher secrecy with respect to scientific research, and lower levels of openness to international scientific interdependence.[25] We can measure relative openness or closure to international scientific interdependence through the degree of state intervention in, or control of, transborder scientific flows and exchanges. These flows include incoming and outbound human resources, knowledge, ideas, and research and experimental equipment. States may be closer to the "liberal" end if they impose relatively few restrictions on exchanges in either direction. If they apply tight controls on such movements they can be characterized as more "mercantilistic" or nationalistic with regard to scientific research and outputs.

---

maintaining scientific productivity. Himmler and Goring recognized the importance of basic and applied atomic science and advocated the special treatment of scientists.

23. See Frieman in this volume and Suttmeier 1980.

24. Skolnikoff 1977. On the relationship between scientific knowledge and the perception of interdependence see Haas 1975.

25. "Research" throughout this discussion also includes development. For a comprehensive survey of the relationship among science, technology and military R&D see Sapolsky 1977. On economics and national security see Kapstein 1992.

Second, the nature and degree of a state's participation in the international economic system affects the domestic emphasis on research. The latter may be more or less geared to maintaining a broadly competitive industrial basis over a military one.[26] The Federal Republic of Germany and Japan, for instance, rank first and second worldwide in civilian R&D expenditures as a percentage of their gross national product (GNP).[27] Greater involvement in the global economy may also imply higher levels of openness to scientific interdependence. In embracing alternative strategies, national ruling coalitions influence the structure of their scientific communities. If, for instance, global or regional power and security considerations are regarded as paramount in guiding a state's domestic and international priorities, military-related R&D will be likely to employ a significant portion of the scientific community.

These external conditions interact with the domestic characteristics discussed earlier to shape a country's "political economy of science."[28] The latter can be defined as *the set of interacting local and international, state and market processes affecting the demand and production of scientific knowledge, its location, and the associated distribution of costs and benefits*. The political economy of science thus provides the global and national contextual boundaries within which state-scientists relations unfold. In other words, domestic and international structures can influence the size and characteristics of the research community, its engine of growth (university, private, or government-created), disciplinary distribution and diversification, social and geographical concentration, levels of engagement in basic versus applied science, nature of rewards, employment patterns and mobility, and, conceivably, even its internal organization (monolithic, pluralistic) and leadership patterns.

I have dealt with the impact of domestic structures on the political economy of science in the previous section. As to external influences, security considerations seem to precipitate fairly similar responses by significantly different states. Such responses include attempts at self-sufficiency and scientific autarky in areas related to military power. The phenomenon of a military-industrial complex sustained by a scientific infrastructure is far from unique to any single political-economic system. In his chapter on the United States Smith argues that, "national security" was the "efficient cause," in the Aris-

26. Domke (1988), for instance, found that high levels of exports as percentage of GNP are associated with lower levels of conflictual behavior. In others words, "governments of nations that are more involved in foreign trade are less likely to make decisions for war" (137).

27. *Spectrum* (November 1989): 67. In general, less spending on military R&D follows a "diffusion-oriented" science and technology policy, in contrast to the "mission-oriented" model of the United States and France (Gummett and Reppy 1987).

28. For a definition of political economy more generally see Gilpin 1987, 8–11.

totelian sense, for the postwar model. Military-related R&D accounted for over 93 percent of total federal R&D outlays in 1963, over 50 percent in 1979, and 70 percent in 1987. The Pentagon and military contractors employed about one-third of U.S. scientists and engineers.[29] The French *force de frappe* was estimated to employ 50 percent of the country's researchers. The former Soviet Academy of Sciences, and at least one third of the institutes, had contracts with the military, representing 10 and 20 percent of their activities, respectively. Nuclear cities like Chelyabinsk-70 and Arzamas-16—the counterparts of Los Alamos, Hanford, and the like—became the hidden reservoirs of the best scientific talent. Military-related R&D in China accounted for 22 percent of all R&D expenditures in 1973 and, as Frieman notes, attracted large numbers of scientists. Industrializing countries, whether subject to explicit national security threats (such as India and Israel) or not (Brazil and Argentina) "sheltered" scientific disciplines for their perceived impact on national security considerations, particularly as defined by their military establishments. Military projects absorbed over 36 percent of R&D funds in Brazil in 1987, and the overall military effort has attracted 50 percent of R&D funds and a large proportion of Israeli scientists.[30]

Military competition is not the only external influence on the domestic political economy of science. The concern with "the American challenge" in the 1960s led to European efforts to offset the United States's edge in science, technology, and industrial production. More recently, economic competition, particularly among advanced industrial countries, has exacerbated what Gilpin (1968) labeled "scientific Colbertism." Elsewhere Gilpin (1975, 131) reminds us that "science policy in France has been first and foremost a product of France's international position." The primary concern of France's leadership has been its "economic and military position relative to other major powers." More specifically, Gilpin traces state strategies vis-à-vis science in France under the Fifth Republic to the recovery of national *grandeur,* the need to deter the Soviet Union, the attempt to contain U.S. economic and political influence, and the vie for leadership in Western Europe. The result was heavy investments in space and military programs, and R&D contracts to universities and private industry (to awaken them to the external political challenges); in other words, a military-industrial complex. Echoing this emphasis

29. See Winner 1992.

30. In civilian R&D spending as a percentage of GNP, the United States trails Japan, West Germany, France, and Britain (Ronayne 1984, 208). For data on military R&D, see Reppy 1989, and Smith 1990 on the United States; Suttmeier 1980, and Frieman in this volume on China; Gilpin 1968, Baumgartner and Wilsford in this volume on France; Kapur in this volume on India; Low in this volume on Japan; and Steinberg and Schwartzman in this volume on Israel and Brazil, respectively.

on external considerations, Baumgartner and Wilsford argue in this volume that economic competitiveness, arms exports and the maintenance of its international position guide French policy vis-à-vis science, reinforcing a centralized institutional structure. International competitiveness is also Japan's dominant objective, as Low postulates in this volume; the state provides tax incentives for domestic R&D, screens foreign know-how, and devotes 98 percent of R&D allocations to civilian sectors. Smith's chapter refers to the increased [U.S.] support for research (including basic research) under the Reagan administration as linked to the perception that U.S. "leadership in the international marketplace is at stake."[31]

The definition by domestic policy networks of a state's position in the international economy, in other words, influences the state's political economy of science. Thus, a liberal approach emphasizing economic interdependence reinforces openness to scientific interdependence, and at least a theoretical preeminence of market forces, or, in the realm of science, of private corporate research. However, as suggested earlier, state intervention in R&D funding and priorities is not a characteristic of centrally planned economies alone.[32] Mercantilistic policies, often prevalent in countries trying to catch up with the "world frontier" through closure, rely on state power to force march the national development of a scientific discipline, evoking an "infant industry" argument.[33] In extreme cases, a state may place barriers on the international mobility of its own scientists and on the export of the fruit of their endeavors, particularly if such fruits can lead to an increase in the relative power of another state.[34] A state may also protect against incoming foreign scientists and screen incoming knowledge or scientific equipment if these are perceived to be carriers of "subversive" politics embodied in scientific trappings. Such "scientific protectionism" was a basic strategy of totalitarian states—and is still alive in the People's Republic of China, as Frieman describes in this volume. The attempts to integrate the PRC into the global economy in the 1980s was paralleled by a remarkable degree of openness to scientific interdependence during those years. The Tiananmen Square debacle

31. The quote is from Dickson 1988, 16.

32. There is significant variation, for instance, in state intervention in R&D (through financing, patent legislation, and tax incentives), ranging from reluctance to provide extensive funding for industrial innovation (United States), to a more active role (Europe), to a very intense state leadership (Japan and the newly-industrialized countries [NICs]). See Schmandt 1981.

33. As Oshima (1980) notes, mercentilitic objectives can be hidden behind support for scientific research projects, as is the case with Japan's satellite launching program.

34. Such strategy of export controls was at the root of Cocom efforts to prevent the diffusion of scientific and technological capabilities into the Soviet Union and formerly Eastern-bloc countries. Export controls generally obey fears of military or economic competition.

altered the leadership's attitudes toward scientists, resulting in a sudden shift toward closure. The cyclical changes in the Soviet case are documented in detail in Josephson's chapter.[35]

Under liberal models, restrictions on diffusion are, in theory, less stringent, at least in areas without obvious immediate relevance to national security.[36] In practice, however, not only does much research produced in the private sector increasingly remain under protective containment for commercial reasons, but even nonmilitary academic research is at times threatened with restrictions.[37] The perceived decline of U.S. power has fueled arguments favoring such restrictions. The implications of emerging patterns of global economic competition, particularly evident in the area of high technology, for the role of science in science-intensive industries, are far from obvious. Current trends may lead to greater secrecy and restrictions on scientists, but they may also encourage increased cooperation—through multinational operations, for instance—on the basis of mutual commercial benefits. Whether economic rationality or political objectives will prevail is largely contingent on unfolding (and contradictory) processes of globalization, on the one hand, and continued strong state intervention, on the other.

States with a pluralistic market-oriented, nonprotectionist model of scientific-technological exchange are hosts—more frequently and extensively than their counterparts—to a very active network of international and transnational actors. These include multinational corporations in the economic realm and nongovernmental nonprofit as well as intergovernmental organizations such as the International Council of Scientific Unions, Greenpeace, the Pugwash Conference on Science and World Affairs, and others in the professional one.[38] These transnational linkages strengthen the scientific community at home, while decreasing the ability of the state to control scientific exchanges. In a cyclical path of influence, therefore, states which are at the outset less prone to impede free transborder flows, may in turn be weakened

35. See also Vucinich 1984, Badash 1985, and Rabkin 1988 on restrictions on Soviet scientists. Interestingly, according to Rabkin, travel overseas was more open to applied scientists and engineers who could pick up technological advances in the West.

36. The application of the McCarrant Act (Internal Security Act of 1950) and McCarrant-Walter Act (Immigration and Nationality Act of 1952) denied U.S. visa entries to distinguished European scientists.

37. The Reagan administration placed limits on diffusion of research results on superconductivity by precluding foreign nationals from participating in a scientific conference on this topic. Other recent control mechanisms included restrictions on foreign visitors to university campuses, limiting the flow of basic scientific information abroad, prepublication agreements, and tightened classification rules. See Smith in this volume, Reppy 1989, and Goggins 1986.

38. On different types of international scientific and professional associations see Crane 1973 and 1981 and commentaries by James R. Kurth and Henry Teune in the same volume.

in their attempt to impose controls on their scientific communities, by the implications of the very openness they predicate.

State-Scientists Relations: Four Ideal Types

The specific characteristics that the political economy of science imposes on a given scientific community may influence the way in which the latter reacts to various forms of control and inducements. What can we learn about patterns of scientists' behavior in different political-economic contexts? Which are the major issues triggering and sustaining, as well as the modes and institutional means used in, collective action? How does the latter differ between contexts where scientists are integrated in state programs and those where they are marginalized?[39]

The answers to some of these questions point to four models—analytic constructs more than empirical descriptions—of state-scientists interaction. From the first to the last, the ratio of autonomy to accountability of scientists decreases. In each model, the tension between the ethos of science and that of political institutions takes a different dimension.[40] A first model—one I label *happy convergence*—assumes a high degree of consensus between state structures and the interests and aspirations of scientists.[41] The latter are not only provided with internal freedom of inquiry but also with freedom of mobility across reward structures (public and private) and the recognition of intellectual property rights. As with any other allocation of resources, the provision of funds for basic scientific research can lead to grievances. These, however, tend to be resolved through more or less institutionalized channels. State-scientists interactions tends to take place in a pluralistic, decentralized environment of competing funding agencies and contending political institutions. High levels of openness to scientific interdependence—domestically, between scientists and the state, and transnationally—reinforce resistance to state intervention in, or control of, the direction of scientific research. The private sector attracts a significant share of research scientists. Happy convergence

39. Among mechanisms of integration or cooptation are effective participation in policy-making, including, but not limited to, science and technology policy; generous funding of scientific activities and other material rewards; creating socioeconomic and cultural demand for scientific expertise; and tolerating a reasonable degree of professional autonomy. Marginalization can be an active strategy, through repression, exile, and forced captivity, or it can be a passive one, through suspicion, indifference, and attitudinal hostility.

40. The "ethos of science" are norms governing scientific research, such as universality (of truth-claims), common ownership of goods and products, and disinterestedness. See Merton 1973. On "autonomy" in science see Cozzens 1990.

41. I have borrowed the term *happy convergence* from the literature on the relationship between economic forces and the state. See Krasner 1978 and Jacobsen and Hofhansel 1984.

often implies a relatively passive scientific community not particularly salient, in political terms, among other societal groups.

Greater happy convergence arguably tends to flourish among pluralist systems of several varieties. The United States became in the 1950s the model for the "scientific state" whose major characteristics, administrative and procedural, included a "partnership among government, industry, and the universities; the growth of nonprofit study and consulting organizations providing services to the government; and the high degree of mobility of professionals, many of them with scientific and engineering backgrounds, among different organizations in the private and public sectors."[42] Smith's chapter in this volume describes this postwar bargain that accepted the autonomy of the scientific enterprise while ensuring its contributions to national security and economic strength.[43] Yet such a model did not exclude state-imposed controls on choices of research topic and on the dissemination of scientific information. The visible hands of the State, Commerce, and Defense Departments—particularly during the Reagan administration—were extended into new definitions of security classification, export controls, and restrictions on foreign scholars.[44] Moreover, there were some discontinuities in happy convergence, particularly with respect to nuclear policy, arms control, the Vietnam War, and environmental issues.

The Federal Republic of Germany was closer, perhaps, to the U.S. model of "happy convergence," than to the more stable French and Japanese ones analyzed in this volume. Baumgartner and Wilsford label the French statist variant "a model of fusion." Low embeds the Japanese model within the structures of "reciprocal consent." In the Israeli case, Steinberg describes a consensual pattern evolving out of a shared understanding that science was to play a special role in offsetting the external disadvantages of a small, developing, and embattled state. Kapur's analysis of India yields more than convergence; scientists *captured* political power by creating a state within the Indian state.

A second model reflects an underlying tension between states and scientists. Such tension may originate in moderate dissatisfaction with systems of domestic controls or with the imposition of constraints on international interactions. The scientific community responds with *passive resistance*. For the most part, this pattern of accommodation appears to have been prevalent, according to Josephson and Pfetsch, during "benevolent" (Lenin, Kruschev,

42. Schmandt 1981, 64.

43. As Mukerji (1989) argues, the state not only "accepts" scientific autonomy, it blesses it, because "independence" or detachment renders science more objective, and, therefore, more authoritative. The state can thus use science as a resource with legitimacy. On the U.S. model see also Price 1965, Kuehn and Porter 1981, Averch 1985, and Dickson 1988.

44. Goggins 1986; Dickson 1988.

Gorbachev) periods of Soviet history and under Nazi rule in Germany. The terms of the science pact in this variant reduce scientific autonomy and the freedom to interact with the international community to a more or less "tolerable" degree from the point of view of the conduct of scientific inquiry.[45] The high relevance of security concerns and relatively low levels of involvement in the global economy tend to skew the scientific community toward military-related research. The state controls and funds most scientific research.

This latent, albeit passive, tension falls short of the heightened levels of conflict under the third model—one characterized by *ritual confrontations*. Here there is a more or less routine and expected expression of grievances on the part of the scientific community, on the one hand, and a built-in animosity toward them among state officials, on the other. Scientists may blame their exclusion from state priorities on external dependency on metropolitan economies. State-scientists relations among many authoritarian regimes in the industrializing world, such as Brazil and Argentina in the 1970s, provide examples for this model as do, according to Low, fractions of the Japanese scientific community.[46] The degree of planning of scientific research—even among market-oriented types—varies according to the state's relative involvement in the international security and economic areas. The absence of political pluralism, particularly when combined with high external security considerations, results in a highly restrictive political context of scientific inquiry, as in the People's Republic of China and postrevolutionary Iran.

This pattern of ritual confrontation can slide into a more extreme adversarial form, and lead to *deadly encounters* of the kind evoked by Stalinism, the Chinese Cultural Revolution, and some authoritarian (mainly military) developing countries. Political accountability replaces any vestiges of scientific autonomy in this model. The practice of scientific "protectionism" is the norm and theoretical research is often detracted as "antisocial." Moreover, scientists become specific targets of persecution and physical annihilation. Entire generations of scientists and whole scientific disciplines have disappeared in this fashion, either through internal physical elimination in gulags or through enforced exile.[47]

These ideal types, of course, can coexist in the real world. Although it is generally possible to identify a prevailing pattern, different subgroups within the scientific community may relate differently to the same political regime,

45. Although the Nazi regime refrained, to some extent, from imposing a Nationalist-Socialist litmus test for scientific activities at the individual level, it did so institutionally. The Prussian Academy, for instance, became a visible instrument of ideological indoctrination.

46. Solingen 1993b.

47. In the PRC only 53 institutes (and 13,000 personnel) remained by 1973—in the aftermath of the Cultural Revolution—from a high of 106 research institutes and over 22,000 scientific and technical personnel in 1965 (Orleans 1980).

by providing support, withholding it, or actively fighting it. Encounters mobilizing significant numbers of scientists—as in the Unites States in 1945–46 (Atomic Energy Bill), 1949–50 (hydrogen bomb), 1963 (Limited Test Ban), and 1967–69 (Antiballistic Missile Treaty (ABM)—may break the normalcy of happy convergence. Kapur's account of the Indian case and Low's chapter on Japan reveal a pattern of happy convergence for insider scientists, and one of ritual confrontation over nuclear policy. Passive resistance under benevolent phases of authoritarian regimes may temporarily replace a tradition of deadly encounters, as in the post-Stalinist period in the former Soviet Union. A similar transition took place in the People's Republic of China following attempts to integrate the country into the global economy in the 1980s, and to increase scientific interdependence. As Frieman argues, a retreat toward greater closure followed the Tiananmen Square debacle. Ritual confrontation may develop out of what Schwartzman labels "benign neglect" in his chapter on Brazil. Yet political liberalization and heightened integration into the world economy have reversed old cycles of confrontation and neglect in quasi-pluralist developing states like South Korea and Taiwan. As some of the chapters suggest such dynamics can be traced to changes in the definition of external security and economic interdependence, and to their related internal political expressions.

The Scientists' Dilemma: The Case of Physicists

> Physicists have vested interests in national policy. Physicists are solicited to advise and consent to policies of the government. Physicists' support is sought by opponents of the government. The "production" of physicists . . . and the wages of physicists are strongly [I say overwhelmingly] influenced by government policy. . . . To take position on public issues is not a matter of choice, it is essential to our being physicists. (Fox 1968, 13–15).

There are robust justifications for the special emphasis on physics by the authors of this volume. First, physics can be considered a classical case[48] of mutual interdependence (and mutual intrusion) between the state and a scientific discipline, particularly in connection to modern technologies. It is more dependent on state support than most other sciences, because of the high costs of research and the unpredictability of immediate results.[49] Second, indus-

48. The subtitle borrows from Drell 1987, 195.

49. Among the natural sciences, physics is in theory the "pure science" par excellence, although solid state, plasma, high energy, and astrophysics have obvious industrial applicability in the areas of electronics, materials, nuclear fusion, and space research. All these are central to so-called frontier sciences.

trialized and developing states of disparate political-economic varieties have bestowed important social functions on physicists, considered to be "at the heart of the ideology of expertise."[50] Third, as a result of their link to nuclear energy and atomic weapons, the political visibility of physicists grew in settings as disparate as those of the United States, the former Soviet Union, India, Brazil, Germany, and the People's Republic of China. Strategic policy, with its domestic and international ramifications regarding the content and context of research, thus became a classical arena of clashes between scientific groups and state priorities. For those actively opposing the development of nuclear deterrence as a doctrine, the "original sin" imparted a special case of "social responsibility."

The political role of physicists in the United States, particularly in the area of national security, has been amply documented in numerous studies.[51] Physicists were dominant in the debates over civilian control of atomic energy, the establishment of the National Science Foundation, the ABM and the Strategic Defense Initiative (SDI).[52] The Committee for Human Rights in the Soviet Union was established in 1970 by physicists Andrei Sakharov, Andrei Tverdokhlebov, and Valery Chalidze.[53] Sakharov, Sagdeyev, Velikhov, Alferov, and others then went on to play central roles in the glasnost era. Physicists like Homi Bhabha and Abdus Salam were important in shaping the relationship between the Indian and Pakistani states and their scientific communities. In the People's Republic of China, the case of astrophysicists Fang Lizhi and Li Shu-Xian was at the heart of the aftermath of the Tiananmen Square massacre, but there were earlier encounters between physicists and the political leadership.[54] Brazilian physicists played an important role in galvanizing the political opposition to military rule, around the industrial-technological characteristics of the country's nuclear program in the 1970s. West German physicists, particularly at universities, were a "decisive political resource" in the antinuclear debates and demonstrations related both to safety

50. Levy-Leblond 1976.

51. These are, however, overwhelmingly concerned with scientific advisers and with the community's prominent members. Physicists were prominent in the Federation of Atomic Scientists, in the 1969 MIT movement that resulted in the Union of Concerned Scientists, in the Presidential Science Advisory Committee, in U.S. Pugwash, and within Scientists and Engineers for Social and Political Action (Schwartz 1971; Nelkin 1972).

52. Cahn 1974.

53. One of the Committee's experts was physicist Boris Tsukerman (Lipset and Dobson 1972).

54. The scientific leadership that replaced those purged in the Academy in 1988 included Moscow-educated hard-liner Zhou Guangzhao, also a theoretical physicist. While many among the older generation of foreign (Western) educated scientists were nuclear physicists, the Soviet-trained generation of the 1950s had a more applied (industrial and engineering) orientation.

and to international security.[55] Their French and Japanese counterparts, instead, were, for the most part, a much more reluctant opposition, as one could expect in a statist model of the kind carefully outlined by Low and Baumgartner and Wilsford. Italian nuclear physicists dominated science policy and acted as spokesmen for the entire science community.[56] The different chapters in the volume suggest that in most cases physicists have rallied, with differing emphasis, around four core political issues: nuclear energy (in its civilian and military dimensions), science policy, human rights, and socioeconomic development.[57]

In Baumgartner and Wilsford's account, nuclear energy reflected the quintessential model of happy convergence in France.[58] They trace the roots of this consensual pattern to widely accepted political priorities regarding the country's international position: energy independence, national prestige, economic competitiveness. Even the nuclear weapons program never created the levels of controversy evident in the ABM or SDI debates in the United States. The Israeli case, according to Steinberg, is another instance of a depoliticized nuclear program.[59] In India, Kapur describes an active scientific elite in nuclear policy that steered the state toward military objectives and away from multilateralism—but a passive, mostly supportive rank and file. The fact that all three cases—France, Israel, and India—are democratic, secondary or regional powers with high potential for involvement in external conflicts and no protection from a superpower's nuclear umbrella is suggestive. Where controversy reached heightened levels of politicization, as in the United States and Brazil, physicists played an important role in catalyzing broader political movements, addressing nuclear strategy in the United States and democratization in Brazil. Low reviews the influence of nuclear physicists in Japan's Science Council and in the politicization of nuclear policy, while emphasizing the largely successful cooptation of most of this community through the privatization of conflict by science-policy bureaucracies.

In the post-Cold War era the relationship between physicists and classical national security concerns is likely to be transformed by the new "high politics" of economic and environmental issues, and by the growing erosion

55. However, many have also publicly endorsed nuclear power (Nelkin and Pollak 1981). On physicists and scientific internationalism in the Weimar Republic see Forman 1973.

56. Cambrosio 1985.

57. For a comparative analysis of political structures and antinuclear movements see Kitschelt 1986.

58. See also Scheinman 1965. There was some opposition among French scientists who signed petitions in the 1970s critical of nuclear power. They never succeeded, however, in coalescing a meaningful public opposition (see Baumgartner and Wilsford in this volume).

59. Of course, as Sanford Lakoff pointed out during the UCLA workshop (January 1990) a depoliticized issue may in fact be a highly political one, where there is overwhelming consensus—and therefore little public debate.

of national boundaries, in scientific research as well as in every other sphere of human activity.

An Obituary to the State-Centered Political Economy of Science?

Emerging global trends are bringing about a possible convergence of models of state-scientists relations toward the end of the twentieth century.

First, military-industrial complexes and their scientific cadres are likely to decline as a consequence of economic conversion to a civilian economy. This demilitarization of science and technology will grow in tandem with the need to maintain global economic competitiveness, and with the strengthened role of international institutions in solving international conflict.[60] The ability of scientists to exploit actual or perceived external threats and to exert domestic political pressures—a central theme in Kapur's analysis of India—may decline concomitantly. The effects of the decline in military competition are evident in a recent U.S. government contract to hire Russian scientists at the Kurchatov Institute of Atomic Energy in Moscow to advance fusion research within the U.S. Department of Energy.[61] Even leaders of the Pentagon's Strategic Defense Initiative ("Star Wars") are now pressing to purchase extant Russian scientific knowledge in space technologies. The thawing of the Cold War has led to one of the most profound transformations in the political economy of science in both the United States and the former Soviet Union.[62]

Second, the globalization of markets, production, finance, and R&D activities increasingly strain the definition of "national" scientific (and other) communities. According to Baumgartner and Wilsford, even France now recognizes the shortcomings of a state-led research infrastructure. The "borderless labs" of multinational corporations have prevailed over states' attempts to concentrate R&D activities on their soil. The search for economic competitiveness spearheads the worldwide diffusion of research activities. Increased economic competition on a global scale, in turn, heightens the share of "private good" research output as well as private sector influence over the direction of research priorities, directly, through university–private sector partnerships, or indirectly, through corporate influence over governmental policy.

60. See Jacobsen 1984 for a pioneering analysis of the impact of international organizations on a future world order.

61. *New York Times,* 3/1/92 and 3/6/92. Similar transactions include the Pentagon's purchase of a Russian nuclear reactor for space power and of plutonium-238 to build nuclear batteries for a new generation of deep-space probes.

62. In the United States hundreds of thousands of scientists and engineers are adjusting to economic conversion after thirty years (and over $1 trillion) of R&D activities in nuclear and other weaponry (*New York Times,* 2/5/92).

Even university–private industry cooperation is growing increasingly transnational; investments by technology-intensive private Japanese firms seek the world-class leadership of U.S. universities in fundamental research.[63]

Third, the expansion of liberal democracy worldwide foreshadows a more homogenous set of political structures than was the case for most of the twentieth century. There is more than openness to scientific interdependence among the new states of the former Soviet Union and Eastern Europe. Hundreds of scientists from these countries are being hired by other countries; paradoxically, unlike others earlier in this century, this brain drain follows liberalization rather than political persecution. These developments are not free of potential dangers. Aggressive or pariah states in search of scientific and technical talent may reap the benefits of an abundant supply of underemployed scientists, many of whom were formerly employed by the military-industrial complex. This danger led to the idea of an internationally funded science and technology center in Moscow to enable scientists to pursue peaceful scientific research. A market failure—an outcome of the difficult economic transitions that the former Soviet Union is undergoing—led to a creative, albeit partial, institutional solution.

Fourth, large-scale cooperative science programs like the space station and European Center on Nuclear Research (CERN) strengthen supranational and/or transnational allegiances among scientific groups—and increasingly defy a purely national research focus.[64] The superconducting supercollider began as a national program, but its financial requirements created growing pressure to internationalize it, particularly via Japanese funding. Even commercially oriented–European Community-funded research projects like Esprit compel cooperative efforts among European firms.

Finally, global environmental challenges compel planetary-centered scientific activities that, in turn, strengthen a truly transnational community with increasing power to influence their respective domestic political contexts. At the same time, private scientific networks, such as the Federation of American Scientists and the Natural Resources Defense Council, as well as their Russian counterparts, are beginning to play truly critical roles in steering cooperation around issues of disarmament and economic conversion.

---

63. See, for instance, the investments by Japanese firms in Princeton, New Jersey, and Palo Alto, California. Many foreign firms participate in "industrial-liaison" programs with top U.S. universities, which grants them early access to mostly federally funded research results. While some decry those developments, others—such as the president of the National Academy of Sciences, Frank Press, and the former director of the National Science Foundation, Ernst Bloch—urge foreign beneficiaries to increase their support for the U.S. research base (*The New York Times,* 24 May 1991: C2).

64. For a pioneering study of European Center on Nuclear Research and other European labs see Teich 1974.

A Methodological Note and Suggestions for Future Research

The underlying assumption of this volume is that domestic structures and international context influence the nature and characteristics of national scientific communities, and their relationship to the state. In particular, such relations are affected by varying responses to the challenges of external conflicts and international economic competitiveness. A more or less common response in the past was to embrace "national-security" programs like nuclear and space technology, widely regarded as essential for the modernization of industrial systems. This was as true for France and India as it was for Brazil and the People's Republic of China, cutting across political-economic systems.

Beyond these general uniformities, there are other ways in which changes in state strategies and/or in the international context affect state-scientists relations. For instance, increased developing country participation in the international economy broadens the demand for domestic scientists, as the experiences of Japan, Taiwan, and South Korea reveal. These changes generally reverse old cycles of confrontation and neglect between scientists and the state. Traditionally, scientists in these countries often blamed the state's strategy of integration into the international division of labor for their own economic marginalization—a result of deficient private sector demand for scientists. In newly industrializing countries like South Korea and Taiwan, however, an invigorated and internationally-competitive position contributed to a "reverse brain drain" from higher institutions in the United States and Europe to new science centers in what used to be labelled the world's "periphery."[65] Another example is how a decreased concern with external military threats tends to soften "ritual confrontations" or "deadly encounters" within noncompetitive centrally planned regimes, as Josephson suggests of the Khrushchev era. Heightened levels of external conflict, instead, may strain happy convergence; India's wars with China and Pakistan had such an effect, as Kapur emphasizes. Yet another example points to the impact of greater closure to scientific interdependence on the radicalization of passive resistance; the Cultural Revolution in the People's Republic of China—as discussed by Frieman—and the Stalinist period in the Soviet Union had such an effect.

From a methodological perspective, the selection of countries aims mostly at exploring variation across systems.[66] Some chapters attempt to

---

65. On a reversal of the brain drain from Taiwan and Korea see *Studies in Comparative International Development* 27, no. 1 (Spring 1992).

66. On variation-finding comparisons see Tilly 1984.

generalize from their own case studies while others pause mostly on the uniqueness of the country's experience. Despite the effort in this introduction to universalize from disparate cases and search for some conceptual unity, the chapters differ in their choice of method and of preferred level of analysis. Some are more historically oriented than others. Yet the focal interest on the relationship between scientists and the state, and on its domestic and external contexts, remains. *Gaiatsu* (or "foreign pressures") argues Low, influenced the nature of political support for science in Japan. *Zhdanovschina* in the Soviet Union and the Cultural Revolution in the People's Republic of China attacked "cosmopolitanism" and threatened to cripple scientific research. Internationally recognized *grandeur* fueled French science policy.

A concrete example of cleavages regarding the appropriate level of analysis is the lively debate within the scholarly community over the sources of glasnost and perestroika.[67] There are those who argue that scientists and technical elites were instrumental in swaying the old regime toward reform. Others maintain that changes in domestic political relations ought to be interpreted in light of the declining international position of the Soviet Union. A weakened, uncompetitive economy and polity necessitated domestic reform, including a new social pact between scientists and the state.

Some papers emphasize ideology as linking structural features (such as a state's international position) on the one hand, and scientists' attitudes on the other. Kapur, for instance, regards India's colonial experience as very powerful in shaping the independent state's strategies on science. Steinberg points out that Zionist ideology perceived science and technology as a source of international political recognition and leverage. Pfetsch argues that, following every military defeat (1808, 1919, late 1940s), the German scientific community committed itself to a new institutional and professional ethos.

Some authors in this volume (Josephson, Kapur, Low, and Steinberg, in particular) elaborate the role of elite scientists, both "insiders" and "outsiders," and of certain political leaders in steering the state's strategy toward science policy and the scientific community.[68] This preference for an individual and elite level of analysis at times complements (Kapur, Low)—and, at others, challenges (Schwartzman)—the more general statements about the role of international and domestic structures. Yet the latter two could arguably have causal precedence: changes in the political economy of science may help propel certain scientific advisors or ideologies into the highest chambers of politics.[69] In other words, structural characteristics may be a useful and perhaps necessary, point of departure, albeit not always sufficient in providing a

67. See Dickson 1988.

68. On "insiders" versus "outsiders" see Primack and Von Hippel 1974.

69. India's independence from British colonial rule and its perceptions of regional threat had precisely such an effect. The external context helped certain domestic coalitions to prevail.

full account of the evolution of different patterns.[70] A focus on bureaucratic and personality factors can shed light on ways in which international processes intersect with, or become part of, domestic political debates.

The framework thus provides a baseline for the formulation of more specific hypotheses. Further research along these lines will need to specify the interaction between domestic regimes and the international context. It may well be, for instance, that, where the former are relatively stable (i.e., longevous), external challenges may have a more powerful influence on changes in domestic political and economic relations.[71] This seems to be the case with the United States where the perception of a declining hegemony transformed, in many ways, the nature of what Smith labels "the science pact."

The authors suggest three other potential lines of inquiry. The first is the need to explore the possible relations between state strength, the strength of market forces bearing on scientific activities, and the political strength of scientific communities. For instance, strong states like France and Israel seem to host relatively weak or at least politically inactive, scientific communities.[72] Instead, these communities are stronger among weak states like the United States and Britain.[73] A second phenomenon worth exploring are the sources of possible internal contradictions within the state (around bureaucratic agencies or personalities) regarding science and scientists, which may prevent them from formulating a more or less coherent strategy. Schwartzman, Low, and Kapur analyze this phenomenon in Brazil, Japan, and India. Third, the relationship between the structural political and economic context and prevailing ideologies of scientific communities remains to be explored.[74]

We hope this volume's eclectic intellectual sources and methods will spark the interest of students of science, technology, society, and international

---

70. In many cases there may be a need to take into account what Norton Wise defined, at the 1990 UCLA workshop, as long-term cultural traditions and short-term historical contingencies. Pfetsch's chapter in this volume discusses cultural traditions in German science.

71. Other linkages between external and internal contexts highlight the primacy of foreign over domestic policy, and their respective impact on attitudes vis-à-vis the scientific community. As Roman Kolkowicz suggested at the 1990 workshop, where the foreign policy agenda is primordial, ideological control and diminished autonomy for scientists is the norm (e.g., Stalin and Brezhnev). Where domestic restructuring is a top priority (e.g., Kruschev and Gorbachev), this trend is reversed.

72. The importance of "state strength" was raised by Sandy Lakoff, Ken Conca, and Frank Baumgartner at the workshop. State strength can be measured by the ability to formulate and implement coherent policies that may withstand societal demands.

73. See Crane 1973.

74. Ideological tendencies within scientific communities have been hitherto traced to characteristics of the research environment, such as pure versus applied science (Barber 1952; Lakoff 1977), the socialization into the "ethos of sciences" (Merton 1967; Hagstrom 1965; Kuhn 1962), and disciplinary categories.

relations, of researchers in the politics and history of science and in the political sociology of professions, of scholars and practitioners of science policy, and of natural scientists interested in the social and political context of scientific research. All these disciplines converge in exploring the many ways in which states exploit scientific knowledge as a venue to power and wealth.

REFERENCES

Averch, Harvey A. 1985. *A Strategic Analysis of Science and Technology Policy.* Baltimore: Johns Hopkins University Press.

Badash, Lawrence. 1985. *Kapitza, Rutherford and the Kremlin.* New Haven: Yale University Press.

Barber, Bernard. 1952. *Science and the Social Order.* Glencoe, IL: The Free Press.

Beyerchen, Alan D. 1977. *Scientists under Hitler.* New Haven: Yale University Press.

Ben-David, Joseph. 1971. *The Scientist's Role in Society.* Englewood Cliffs, NJ: Prentice Hall.

Ben-David, Joseph. 1978. "Emergency of National Traditions in the Sociology of Science—The United States and Great Britain." In J. Gaston, ed., *Sociology of Science,* San Francisco: Jossey-Bass. 197–218.

Brooks, Harvey. 1980. "Technology, Evolution, and Purpose," *Daedalus* 109 (Winter): 65–81.

Brooks, Harvey. 1981. "Note on some Issues on Technology and National Defense." *Daedalus,* Winter, 129–36.

Cahn, Anne H. 1974. "American Scientists and the ABM: A Case Study in Controversy." In Albert H. Teich, ed., *Scientists and Public Affairs,* 41–120. Cambridge, MA: MIT Press, 1974.

Cambrosio, Alberto. 1985. "The Dominance of Nuclear Physics in Italian Science Policy." *Minerva* 23, no. 4 (Winter): 464–84.

Clark, Norman. 1987. "Similarities and Differences between Scientific and Technological Paradigms." *Futures* February, 26–42.

Cozzens, Susan E. 1990. "Autonomy and Power in Science." In Cozzens and Gieryn, 164–84.

Cozzens, Susan E. and Thomas F. Gieryn, eds. 1990. *Theories of Science in Society.* Bloomington, IN: Indiana University Press.

Crane, Diana. 1973. "Transnational Networks in Basic Science." In R. O. Keohane and Joseph S. Nye, Jr., eds., *Transnational Relations and World Politics,* 235–51. Cambridge, MA: Harvard University Press.

Crane, Diana. 1981. "Alternative Models of ISPAs." In William M. Evan, ed., *Knowledge and Power in a Global Society,* 29–48. London: Sage Publications.

Dahl, Robert A., and Charles E. Lindblom. 1953. *Politics, Economics, and Welfare.* New York: Harper and Brothers.

Dedijer, Stevan. 1968. "Underdeveloped Science in Underdeveloped Countries." In Edward Shils, ed., *Criteria for Scientific Development: Public Policy and National Goals.* Cambridge, MA: MIT Press.

Dickson, David. 1988. *The New Politics of Science*. New York: Pantheon Books.

Domke, William K. 1988. *War and the Changing Global System*. New Haven: Yale University Press.

Drell, Sidney D. 1987. "The Scientists' Dilemma." In Robert J. Myers, ed., *International Ethics in the Nuclear Age*, 195–210. Boston: University Press of America.

Eckstein, Harry, ed. 1964. *Internal War: Problems and Approaches*. New York: Free Press.

Eckstein, Harry. 1975. "Case Study and Theory in Political Science." In Fred Greenstein and Nelson Polsby, ed., *Handbook of Political Science* Vol. 7. Reading, Mass." Addison-Wesley, 79–138.

Ezrahi, Yaron. 1990. *The Descent of Icarus: Science and the Transformation of Contemporary Democracy*. Cambridge, MA: Harvard University Press.

Forman, Paul. 1973. "Scientific Internationalism and the Weimar Physicists: The Ideology and Its Manipulation in Germany after World War I." *Isis* 64, no. 222 (June): 151–80.

Fortescue, Stephen. 1986. *The Communist Party and Soviet Science*. London: Macmillan.

Fox, Herbert L. 1968. Letter to *Physics Today* (March): 13–15.

George, Alexander L. 1979. "Case Study and Theory Development: The Method of Structured, Focused Comparison." In Paul Gordon Lauren, ed., *Diplomacy—New Approaches in History, Theory, and Policy*, 43–68. New York: Free Press.

Gilpin, Robert G. 1962. *American Scientists and Nuclear Weapons Policy*. Princeton, NJ: Princeton University Press.

Gilpin, Robert G. 1968. *France in the Age of the Scientific State*. Princeton, NJ: Princeton University Press.

Gilpin, Robert G. 1975. "Science, Technology, and French Independence." In D. Long and C. Wright, 110–32.

Gilpin, Robert G. and Christopher Wright, eds. 1964. *Scientists and National Policy-Making*. New York: Columbia University Press.

Goggins, Malcolm L., ed. 1986. *Governing Society and Technology in a Democracy*. Knoxville, TN: The University of Tennessee Press.

Golden, William T., ed. 1988. *Science and Technology Advice to the President, Congress, and Judiciary*. New York: Pergamon.

Gourevitch, Peter A. 1978. "The Second Image Reversed: The International Sources of Domestic Politics." *International Organization* 32:881–911.

Gowing, Margaret, assisted by Lorna Arnold. 1974. *Independence and Deterrence Britain and Atomic Energy, 1945–1952*. London: Macmillan.

Graham, Loren R. 1967. *The Soviet Academy of Sciences and the Communist Party 1927–1932*. Princeton, NJ: Princeton University Press.

Graham, Loren R. 1972. *Science and Philosophy in the Soviet Union*. New York: Knopf.

Graham, Loren R. 1977. "The Place of the Academy of Science System in the Overall Organization of Soviet Union." In John Thomas and U. M. Kruse-Vaucience, eds., *Soviet Science and Technology*, 44–58. Washington, DC: George Washington University Press.

Gummett, Philip, and Judith Reppy, eds. 1987. *The Relations between Defense and Civil Technologies*. Dordrecht: Klewer.

Haas, Ernst B. 1975. "Is there a hole in the whole? Knowledge, technology, interdependence, and the construction of international regimes." *International Organization* 29:827–76.

Haas, Ernst B., Mary Pat Williams, and Don Babai. 1977. *Scientists and World Order.* Berkeley: University of California Press.

Haberer, Joseph. 1969. *Politics and the Community of Science.* New York: Van Nostrand Reinhold.

Habermas, J. 1971. *Towards a Rational Society.* London: Heinemann.

Hagstrom, Warren O. 1965. *The Scientific Community.* Basic Books: New York.

Hirsch, Walter. 1974. *The Autonomy of Science.* In Restivo and Vanderpool, 144–57.

Holloway, David. 1970. "Scientific Truth and Political Authority in the Soviet Union." *Government and Opposition* (Summer): 345–67.

Horkheimer, M., and T. W. Adorno. 1973. *Dialectic of Enlightenment.* London: Allen Lane.

Jacobsen, J. K., and C. Hofhansel. 1984. "Safeguards and Profits: Civilian Nuclear Exports, Neo-Marxism, and the Statist Approach." *International Studies Quarterly* 18: 195–218.

Jacobson, Harold K. 1984. *Networks of Interdependence—International Organizations and the Global Political System.* New York: Knopf.

Jacobson, Harold K., and Eric Stein. 1966. *Diplomats, Scientists, and Politicians—The United States and the Nuclear Test Ban Negotiations.* Ann Arbor: University of Michigan Press.

Jasanoff, Shelia. 1990. *The Fifth Branch/Science Advisors as Policymakers.* Cambridge, MA: Harvard University Press.

Kapstein, Ethan B. 1992. *The Political Economy of National Security.* New York: McGraw-Hill.

Katz, James E. 1982. "Scientists, Government and Nuclear Power." In Everett Katz and Marivah Onkar S. *Nuclear Power in Developing Countries,* 55–75. Lexington, MA: Lexington Books.

Katz, James E. 1986. "Factors Affecting Military Scientific Research in the Third World." In James E. Katz, ed., *The Implications of Third World Military Industrialization: Sowing the Serpents' Teeth,* 293–304. Lexington, MA: Lexington Books.

Kevles, Daniel J. 1978. *The Physicists: The History of a Scientific Community in Modern America.* New York: Knopf.

Kitschelt, Herbert P. 1986. "Political Opportunity Structures and Political Protest: Anti-Nuclear Movements in Four Democracies." *British Journal of Political Science* 16, pt 1 (January): 57–85.

Kneen, Peter. 1984. *Soviet Scientists and the State.* London: Macmillan.

Kornhauser, William. 1962. *Scientists in Industry—Conflict and Accommodation.* Berkeley: University of California Press.

Kramish, Arnold. 1959. *Atomic Energy in the Soviet Union.* Stanford: Stanford University Press.

Krasner, Stephen D. 1978. *Defending the National Interest.* Princeton, NJ: Princeton University Press.

Kuehn, T. J., and A. L. Porter, eds. 1981. *Science, Technology, and National Policy.* Ithaca, NY: Cornell University Press.

Kuhn, Thomas S. 1962. *The Structure of Scientific Revolutions*. Chicago: University of Chicago Press.

Kurth, James R. 1981. "Commentary." In William M. Evan, ed., *Knowledge and Power in a Global Society*, 69–72. Beverly Hills: Sage.

Lakoff, Sanford A., ed. 1977. *Knowledge and Power: Essays on Science and Government*. New York: Free Press.

Landes, David S. 1969. *The Unbound Prometheus*. Cambridge, MA: Cambridge University Press.

Levy-Leblond, Jean-Marc. 1976. "Ideology of/in Contemporary Physics." In H. Rose and S. Rose, eds., *The Radicalisation of Science*, 136–75. London: Macmillan.

Lindblom, Charles E. 1977. *Politics and Markets—The World's Political-Economic Systems*. New York: Basic Books.

Lipset, S. M., and R. R. Dobson. 1972. "The Intellectual as Critic and Rebel: With Special Reference to the U.S. and the Soviet Union." *Daedalus* 101:137–98.

Long, T. Dixon, and Christopher Wright, eds. 1975. *Science Policies of Industrial Nations*. New York: Praeger.

Marcuse, Herbert. 1964. *One Dimensional Man: Studies in the Ideology of Advanced Industrial Society*. Boston: Beacon Press.

Merton, Robert K. 1967. *Social Theory and Social Structure*. New York: Free Press.

Merton, Robert K. 1973. *The Sociology of Science—Theoretical and Empirical Investigations*. Chicago: University of Chicago Press.

Moravcsik, Michael J. 1968. "Some Practical Suggestions for the Improvement of Science in Developing Countries." In Edward Shils, ed., *Criteria for Scientific Development: Public Policy and National Goals*, 177–86. Cambridge, MA: MIT Press.

Mothe, John de la, and Louis Marc Ducharme, eds. 1990. *Science, Technology, and Free Trade*. London: Pinter.

Mukerji, Chandra. 1989. *A Fragile Power: Scientists and the State*. Princeton, NJ: Princeton University Press.

Nau, Henry R. 1972. "The Practice of Interdependence in the Research and Development Section: Fast Reactor Cooperation in Western Europe." *International Organization* 26, no. 3: 499–526.

Nelkin, Dorothy. 1972. *The University and Military Research: Moral Politics at MIT*. Ithaca, NY: Cornell University Press.

Nelkin, Dorothy, and Michael Pollak. 1981. *The Atom Besieged*. Cambridge, MA: MIT Press.

Nelson, Richard R. 1974. "Less Developed Countries—Technology Transfer and Adaptation: The Role of the Indigenous Science Community." *Economic Development and Cultural Change* 23:61–77.

North, Douglass C. 1981. *Structure and Change in Economic History*. New York: W. W. Norton.

Orleans, Leo A., ed. 1980. *Science in Contemporary China*. Stanford: Stanford University Press.

Oshima, Keichi. 1980. "Technology and Economic Nationalism." In Otto Hieronymi, ed., *The New Economic Nationalism*, 201–9. New York: Praeger.

Paquet, Gilles. 1989. "Science and Technology Policy Under Free Trade." *Technology in Society: An International Journal* 2, no. 2: 221–34.

Polanyi, Michael. 1946. *Science, Faith, and Society.* London: Oxford University Press.

Price, Derek John de Solla. 1963. *Little Science, Big Science.* New York: Columbia University Press.

Price, Don K. 1965. *The Scientific Estate.* Cambridge, MA: Belknap Press, Harvard University Press.

Primack, Joel, and Frank Von Hippel. 1974. *Advice and Dissent: Scientists in the Political Arena.* New York: Basic Books.

Rabkin, Yacov M. 1988. *Science between the Superpowers.* New York: Priority Press.

Reich, Robert. 1991. *The Work of Nations.* New York: Knopf.

Reppy, Judith. 1989. "More for the Military." *Science Policy* (Jan–Feb): 46–48.

Restivo, Sal P. 1974. "The Ideology of Basic Science." In Restivo and Vanderpool, 334–51.

Restivo, Sal P., and C. V. Vanderpool, eds. 1974. *Comparative Studies in Science and Society.* Columbus, OH: Charles E. Merrill.

Ronayne, J. 1984. *Science in Government.* London: Edward Arnold.

Ruggie, John G. 1975. "International Responses to Technology: Concepts and Trends." *International Organization* 29, no. 3: 557–83.

Salomon, Jean Jacques. 1973. *Science and Politics.* London: Macmillan.

Sapolsky, Harvey M. 1977. "Science, Technology, and Military Policy." In Spiegel-Rosing and deSolla Price, 443–71.

Scheinman, Laurence. 1965. *Atomic Energy Policy in France Under the Fourth Republic.* Princeton, NJ: Princeton University Press.

Schmandt, Jurgen. 1981. "Toward a Theory of the Modern State: Administrative vs. Scientific State." In J. S. Szyliowicz, ed., *Technology and International Affairs,* 43–97. New York: Praeger.

Schroeder-Gudehus, Brigitte. 1977. "Science, Technology, and Foreign Policy." In Spiegel-Rosing and deSolla Price, 473–506.

Schwartz, Charles. 1971. "Professional Organization." In Martin Brown, ed., *The Social Responsibility of the Scientist,* 19–35. New York: Free Press.

Shils, Edward A. 1962. "The Autonomy of Science." In B. Barber and W. Hirsch, eds., *The Sociology of Science,* 610–22. New York: Free Press.

Simon, Denis Fred. 1987 "China's Scientists and Technologists in the Post-Mao Era: A Retrospective and Prospective Glimpse." In Merle Goldman, ed., *China's Intellectuals and the State—In Search of a New Relationship,* 129–55. Cambridge, MA: Harvard University Press.

Skocpol, Theda. 1985. "Bringing the State Back In: Strategies of Analysis in Current Research." In Peter B. Evans, D. Rueschemeyer, and T. Skocpol, eds., *Bringing the State Back In,* 3–43. New York: Cambridge University Press.

Skolnikoff, Eugene B. 1967. *Science, Technology, and American Foreign Policy.* Cambridge: MIT Press.

Skolnikoff, Eugene B. 1977. Science Technology, and the International System. In Spiegel-Rosing and deSolla Price, 507–33.

Smith, Bruce L. R. 1990. *American Science Policy since World War II.* Washington, DC: Brookings Institution.

Snow, C. P. 1961. *Science and Government.* Cambridge: Harvard University Press.

Snow, C. P. 1981. *The Physicists*. Boston: Little, Brown & Co.
Solingen, Etel. 1989. "State Forms, International Involvement, and the Political Economy of Science." Typescript.
Solingen, Etel. 1993a. "Between Markets and the State: Scientists in Comparative Perspective." *Comparative Politics* 26, 1: 31–51.
Spiegel-Rosing, Ina. 1993b. *Bargaining in Technology: Industrial Policy and Nuclear Programs in Brazil and Argentina*. Typescript.
Spiegel-Rosing, Ina, and Derek J. deSolla Price, eds. 1977. *Science, Technology, and Society: A Cross-Disciplinary Perspective*. London: Sage Publications.
Stepan, Nancy. 1976. *Beginnings of Brazilian Science: Oswaldo Cruz, Medical Research and Policy, 1890–1920*. New York: Science History Publications.
Storer, Norman W. 1966. *The Social System of Science*. New York: Holt, Rhinehart, and Winston.
Suttmeier, Richard. 1980. *Science, Technology, and China's Drive for Modernization*. Stanford: Hoover Institution Press.
Szyliowicz, Joseph S., ed. 1981. "Technology, The Nation State: An Overview." In Joseph S. Szyliowicz, ed., *Technology and International Affairs*, 1–40. New York: Praeger.
Teich, Albert H. 1974. "Politics and International Laboratories: A Study of Scientists' Attitudes." In Albert Teich, ed., *Scientists and Public Affairs*. 173–235. Cambridge, MA: MIT Press.
Teune, Henry. 1981. "Commentary." In William M. Evan, ed., *Knowledge and Power in a Global Society*, 73–76. Beverly Hills: Sage.
Tilly, Charles. 1984. *Big Structures, Large Processes, Huge Comparisons*. New York: Russell Sage Foundation.
Vessuri, Hebe M. C. 1987. "The Social Study of Science in Latin America." *Social Studies of Science* 17, no. 3 (August): 519–54.
Vucinich, Alexander. 1984. *Empire of Knowledge. The Academy of Sciences of the USSR (1917–1970)*. Berkeley and Los Angeles: University of California Press.
Waltz, Kenneth N. 1959. *Man, the State and War: A Theoretical Analysis*. New York: Columbia University Press.
Winner, Langdon. 1992. "The Culture of Technology." *Technology Review* (February-March): 70.

The United States: The Formation and Breakdown of the Postwar Government-Science Compact

Bruce L. R. Smith

The relations between science and society in the United States are multiple, complex, and interactive. There is no fixed pattern that has prevailed uniformly over recent history. Like science itself, the relations between U.S. science and society have constantly evolved. But they evolve against a cultural background and in a constitutional and institutional context that at least gives the observer some points of reference. Here, I focus on the underlying forces and salient institutional dimensions that have shaped society's interactions with its scientific communities. My primary aim is to show how the classic features of scientific activity—autonomy, disinterestedness, and rationality—affect, and are affected by, the workings of the U.S. political system both domestically and in the light of post–World War II globalization of the U.S. economy.

Science and the Founding Fathers

Science has been closely linked with public affairs since the U.S. nation's birth. The metaphor of marriage partners has been used to describe the relationship between science and U.S. democracy—intimate, mutually dependent, but characterized by frequent quarrels (Price 1965, 171). Even more than other democracies, the United States has been open to scientists' involvement in the workings of the governing system in broad policy areas—in "science *in* policy" as well as "policy *for* science" (Golden 1980; Lakoff 1966; Smith 1992). The United States was the first society to abolish the divine right of kings and to substitute empirically based reason (Madison's "new science of politics") for traditional authority. The Founding Fathers separated religion from the state, though its free exercise was constitutionally protected. The Founding Fathers naturally looked to the Enlightenment for the ideas that would guide public affairs (Ezrahi 1990). Lacking authority legitimated by monarchical succession or religious belief, the governing structure would function through a balance of interests representing the values of rationality, political compromise, and procedural fairness. This model of how government would work was itself an eighteenth-century conception, a construct of reason and empirical analysis applied to human affairs. It was never wholly accepted by all the actors in the political system, just as the modern "consensus" was never fully accepted by all political actors. The anti-Federalists bitterly assailed the ideas advanced by the exponents of practical reason as the guide to sensible policy, but these critics were history's losers (Ezrahi 1990, 108; Smith 1992, 12).

Within the governing system, the motive force or the engine was to be practical reason embodied either in outside social interests ("factions") or in the executive departments that would advance ideas for action. These ideas would be directed toward practical problems and reflect specialized professional interests. Science would be engaged in public affairs through its numerous tributaries and specialized subdisciplines, and thus would reflect partial and limited scientific truths rather than any Universal Truth (too close to the religion banished from public life). Because there would be a cacophony of voices as the various specialized interests clashed, the elected politicians would play the role of arbiters and would promote compromise. The judiciary, capped by the Supreme Court, would ensure that the whole process followed certain norms of procedural fairness (from time to time natural law notions of substantive rights crept in disguised as process values).

U.S. government and society were unusually permeable and receptive to scientific ideas. The United States was, indeed, the ideal breeding ground for a scientific and industrial civilization. Joseph Priestley, the great chemist and dissenting Unitarian clergyman, chose the United States as his land of exile when crowds stoned his Manchester home after he supported the French Revolution. The United States embodied the limited government ideas so congenial to Priestley and to his industrialist friends. Tradition, entrenched privilege, and oppressive government had less scope to block the progress of mankind.

But in another sense the United States had a fatal flaw. The flaw resulted from an excess of its virtues. The United States was so hostile to elitism in any form that populism might impede higher learning and progress toward the upper reaches of scientific attainment—so, at any rate, many foreign observers have thought, beginning with de Tocqueville in the 1830s. The United States, it was feared, would concentrate on the practical and the narrowly utilitarian to the neglect of pure science. Thus, while government applied science and industrial technology might flourish, the U.S. universities—and the basic sciences in general—might well lag behind Europe.

For the early part of the nineteenth century, this simplistic picture could be viewed as a crude approximation of reality. Science as an autonomous enterprise—valued and pursued for its own sake—was not the conception of many early leaders in the United States. Indeed, science was seen as the handmaiden of government and of commercial enterprise. To the framers, all knowledge was ultimately practical. They therefore saw no reason for distinctions between how society and how the scientists themselves would define the aims and purposes of scientific activity.

The framers looked at the world in scientific terms; the governing system they created was a metaphor of the Newtonian universe. Forces were balanced and held in equilibrium by counterforces; movement and dynamism coexisted with stability and orderly change. All were connoisseurs of science, some ardent patrons (Thomas Jefferson), and some world-class scientists (Benjamin Franklin and David Rittenhouse). Moreover, science blended naturally into the workings of government departments (mapping, surveys, the army arsenals). The only exceptions were proposals that clearly exceeded the functions of the federal government of limited and enumerated powers (e.g., Thomas Jefferson's idea of a national university, Charles Pinckney's proposal for aid to the arts and sciences, and, later, Alexander Hamilton's call to transform society from an agrarian to an industrial base). Such powers were among those to be reserved to the states or else denied altogether to government. In all subsequent periods of U.S. history, there was no unanimity of view but instead a disorderly and shifting collision of forces producing a temporary majority view.

The "scientific community" was a misnomer. Until 1815 or so, science in the United States was largely practiced by gentlemen amateurs who were also men of affairs. Science was neither a separate profession nor an identifiable, distinct function within the government bureaus or the infant firms in the early nineteenth century.

The Emergence of the Research System

Stating the matter in shorthand, three notable developments took place in the course of the nineteenth century: (1) science emerged as a distinct, professionalized activity with full-time practitioners; (2) a rough tripartite division of labor formed among scientific institutions in government, industry, and the institutions of higher learning; and (3) within each of these sectors a social division of labor emerged among the various guilds of scientific activity. A stratified and complex community could be identified.

The United States' scientific community took on many of its modern features over the 1846–76 period (Bruce 1987). Science became a professionalized pursuit, no longer a part-time activity. The United States produced scientists through its educational institutions who then pursued careers in research, including basic research. There were linkages between science and technology, between scientists in government and in universities, and between those who generated knowledge and those who commercially applied it (Skinner 1985, 63–100).

As science progressed, the contours of what we would in modern parlance call the *research system* emerged. First, government departments would conduct scientific activity directly related to their missions. This government share of the nation's scientific effort principally was applied research (though such leaders of government technical bureaus as Alexander Dallas Bache and Matthew Fontaine Maury played vitally important roles in the broad development of U.S. science). Second, the emerging universities carried out more fundamental inquiry and trained the scientific manpower (some of the best training in the early period also took place in the government technical bureaus). Third, industry was largely responsible for inventive activity and for the commercialization of technical discoveries. Industry commercialized technology and focused on process technologies that improved the efficiency of manufacture. The system, in brief, consisted of three sets of institutions performing distinct but somewhat overlapping functions.

The institutions had close linkages and patterns of interaction, but as the system matured their respective and distinct identities and roles became more clearly defined. The universities' hold on basic research became firmer; no German Max Planck Society or French National Center for Scientific Research (CNRS) diluted its role as the "home of science" (Wolfle 1972). In contrast to the continental European model, and more akin to that of the United Kingdom, basic research had its firmest roots in the universities rather than in the universities plus a network of research institutes not directly involved with teaching (National Research Council 1982, chap. 13). The highly centralist Academy of Science model that developed in Russia and Eastern Europe with most scientific activity outside of the universities is a third variant in state-science relations whose origins and consequences are beyond the scope of this chapter (see Josephson in this volume).

Except in agriculture, the U.S. government in the nineteenth century did not see itself as responsible for moving technology into the marketplace. Even in agriculture, government-sponsored technical activity did not carry over into the commercial food chain where appropriability (or the assigning of brand name and proprietary rights) takes place. Industry monopolized the "downstream" aspects of technology development and spent more on research than did government or the universities. However, industry did not—and does not—form as dominant a role in the nation's total research effort as does, for example, contemporary industry in Japan or civil industry in Germany.

The financial support of each sector's technical activity was, furthermore, its own responsibility in the nineteenth century. For example, it was not the federal government's job to support university research prior to World War II. This was the responsibility of the universities themselves and of private philanthropy (Bruce 1987; Geiger 1986). The autonomy of university science would be better served this way in the view of university scientists. Nor was it government's responsibility to support technology development except indirectly—for example, through the legal order by resolving ownership (patent) disputes, by establishing technical standards, and by serving in some cases as a customer for industrial goods in the early stages of market development.

Further, as the research system continued to evolve, the government, academic, and industrial sectors became increasingly complex internally. Industry, at the end of the nineteenth and in the early twentieth century, developed the centralized research laboratory. At General Electric (GE), for example, the laboratory under William Whitney worked first on manufacturing problems to prove its usefulness to the company (Dupree 1957, 287–88, 305–23; Reich 1985; Wise 1985). This was not the conception of Charles Steinmetz; he wanted the laboratory to be divorced from operational responsibilities. The laboratory subsequently evolved more closely along the lines originally envisaged by Steinmetz. This produced the advantage of long-range technical thinking on new products and processes, but had the potential disadvantage of making research activities become unrelated to or too distant from the marketplace (Graham 1985; Hounshell and Smith 1988; Wasserman 1985). The autonomy of scientific effort considered desirable to promote the greatest professionalism thus created a technology transfer problem within the same organization.

In like fashion, government research and development efforts in the nineteenth century, as they became more highly professionalized, could often diverge from the needs of the operating officials for whom they were ostensibly conducted. Universities, as organizations set up to promote the creativity of individual professors and students, always managed to generate internal conflicts. The missions of teaching and of research, the respective roles of central administration and of faculty, and disputes over the powers of institutes and of departments were only a few of the sources of conflict within the theoretically self-governing republic of science.

The research system, as the United States moved into the modern period, thus involved three sets of relationships encompassing: (1) science and the economy; (2) science and the government; and (3) science and the educational system. The system as a whole was only loosely integrated; the scientific institutions of government, education, and industry evolved along separate tracks. Within each broad sector there was a further complex division of labor between a technical enterprise propelled by its own inner logic and the wider social goals motivating the end users or consumers of the technologies. The system, however, was loose and disorderly enough so that ideas and individuals could move easily across the institutional boundaries. As the system matured, functional specialization tended to produce greater rigidities and less mobility. Effective linkages between and within scientific institutions became more difficult to establish. But the system "worked" in part because little was expected of it. U.S. society tended to be insular, outlooks and attitudes parochial, and public policy issues and cultural life predominantly domestic in orientation. Capital and ideas could be imported to provide dynamism for the economy when domestic sources were lacking. This "traditional" outlook began to change as the United States was drawn into World War I and as the Great Depression shook the political landscape in the 1930s. Scientists became shaken loose from their traditional conception of aloofness and indifference toward government, and political roles began to emerge (Kuznick 1987). But World War I and the depression were merely the precursors to the dramatic changes that took place with World War II.

The Postwar "Science Pact" in the United States

World War II ushered in the modern era of government-science relations in the United States. The role of science in national security was the "efficient cause" in the Aristotelean sense. Military concerns were important factors in the relations between science and society in the nineteenth-century United States, but the contrast with the dominant role played by international concerns in the twentieth century is striking. The Army, for example, needed mapping, surveying, arsenals for ordnance, standardized gauges, and other technical products and services in the nineteenth century, but the technical efforts mostly originated within the military. The American Civil War led to efforts to enlist the services of nongovernmental scientists and industrialists, including the creation of the National Academy of Sciences to provide scientific advice (Dupree 1957, chap. 7). After the war the military services returned to reliance on their own technical staffs. World War I shocked military leaders into the recognition that the nation was ill-prepared for the emergency, lacking nitrate, glass, and other critical wartime industries and having allowed technical capacities within the military to atrophy. Nevertheless after the war the nation once again drifted back to the prewar neglect of technology.

World War II brought far-reaching and fundamental changes. The advent of nuclear energy, radar, proximity fuses, new medical technology, and other developments convinced military officials that they needed technical capacities beyond their own internal bureaus (which also needed to be significantly expanded and improved). Something akin to permanent mobilization began to develop. But how did this new relationship actually happen? What institutional forms and relationships developed between the scientific community and the government? Scientists, who had been willing to put aside their own concerns during the war, were now anxious to rebuild basic science in the universities. Civilian science had been interrupted and its needs largely neglected. Such traditional values as autonomy, open communication, and disinterestedness had been willingly but temporarily suspended.

A formula was called for to accomplish a number of different purposes. A number of functional requirements could in theory be identified. First, a closer integration of the nation's technical capacities was required. Industry, government, and the universities should be integrated more effectively, and should, in fact, be brought together almost on a permanent mobilization basis to maintain a high state of readiness. The Cold War, which accelerated the nuclear, space, and information revolutions in the initial phase, suggested the need for an alliance between science and the military.

At the same time the autonomy of the scientific enterprise, especially in the universities but also in industry, had to be safeguarded. This was essential if the universities, for example, were to produce scientific personnel of the highest quality to satisfy the needs of the expanding system. Moreover, the "bank" of basic knowledge (a favorite metaphor of Vannevar Bush and other leaders of the scientific community) would have to be replenished. While one could in theory identify such "needs," it was by no means evident how society would actually react in the demobilization period. There were numerous conflicting views within the scientific community and among the politicians, government administrators, media commentators, and others who followed such issues.

In this context the famous 1945 report, *Science—The Endless Frontier,* by Vannevar Bush, who was named wartime director of the Office of Scientific Research and Development, became the focus of the postwar science policy debate. From this debate beginning in the war's late stages, and continuing in the intense maneuverings surrounding the creation of the National Science Foundation and the Atomic Energy Commission, arose what I call the postwar "science pact" referred to by Solingen in her introductory chapter to this volume. This pact or social contract stated the terms of an implicit bargain between the leaders of the scientific community and government officials, politicians, and opinion leaders that shaped public policy for nearly a generation. These developments have been analyzed at length elsewhere and there is no need to plow the ground in detail again (Dupre and Lakoff 1962; Kevles 1990; Smith 1990). But a few salient aspects of the struggle to forge the working relationships between science and society should be noted. The policy debate was complex and many-faceted, with numerous splits within the scientific community as well as between executive branch officials and members of Congress and others. The *Science—The Endless Frontier* report was actually a collection of separate analytical appendices that came to no clear resolution of a number of complex problems. Was the government to support research primarily in the university centers that already had significant research programs? Or should an effort be made to develop wider capacities across the country? Should research support be allocated on a formula basis or on a project basis? What would be the relationship among the different research performers—universities, government labs, and industry? Bush, using the prerogatives as chairman of the review effort, resolved the differences within the scientific community by writing an overall summary of the work of the three panels. This short survey became one of the most influential policy documents in U.S. history and was widely accepted as reflecting the viewpoint of the scientific community.

The opposing view, also marked by internal differences, coalesced around Senator Harley Kilgore (D–WV), who was a maverick New Dealer and a critic of big business and the big universities that he considered to be in an unholy alliance with business. Kilgore saw the major objective of postwar science organization as fostering applied science, with the government laboratories and not the universities as the most important institutional actors. He envisaged an extention of the formula-type, largely-applied research efforts characteristic of agricultural policy to the whole national research effort.

Both Bush and Kilgore as well as their respective allies, however, favored the notion of a large entity that would centralize the major functions of the research system (e.g., allocate and disburse research support, plan the early stages of technology development for potential later adoption by the mission agencies and departments, provide for manpower support and assistance, coordinate the government's technical programs, and other functions). Neither side fully thought through or articulated a position on the relationship of government policy to the private commercialization of technology, in part because of bitter disputes over patent policy. The Bush report dodged the issue by calling for a study of patent issues. By its omission from the debate, civil sector technology development was thus left to the incentives of the marketplace. This was, in effect, a carryover of nineteenth-century attitudes and practices.

It is important to note that the international climate provided the backdrop for the unfolding policy debates. The contribution of science to the war effort enhanced the prestige of the statesmen of science and persuaded the military services to begin planning for the expansion of their own technical bureaus as well as programs of external research support. As the Cold War loomed, the need for action to integrate science fully into military missions became increasingly urgent in the minds of military planners. The international context was the critical factor that drove the whole postwar reappraisal of science and technology policy. The emerging Cold War provided the urgency that made possible the social contract between science and society. In the absence of such background forces, science policy may more naturally fall into the "benign neglect" described by Schwartzman in his analysis of Brazilian science policy in this volume.

The working accommodation between the scientific community and society over the 1945–50 period, which prevailed for nearly a generation, did not, of course, emerge full-blown. Nor was there anything like the constitutional convention that debated the issue in some orderly fashion. A series of protracted debates over the proposed National Science Foundation (Kevles 1977; Price 1954), the three-volume Steelman Report of 1947 (Steelman Report 1947), and the independent initiative taken by the Navy Department (Sapolsky 1990) all played important roles in shaping the evolution of the postwar research system, as did the 1948 election that deposed Senator Kilgore as chairman of a key subcommittee. The story cannot be retold in detail here, with its numerous subplots and personality and policy disputes. But, to a surprising degree, when the dust settled a set of loose understandings had emerged among key political actors, media elites, scientists and science administrators that, following Solingen, may be characterized as the science-society pact. *Science—The Endless Frontier* and the subsequent Steelman Report provided the general framework and rationale for the major concepts. So long as the postwar economic reconstruction and defense posture of the West against the Soviet thrust remained on track the framework of science policy received general acceptance, even though there were critics of the formula from the start.

For present purposes it suffices to note the key elements of the bargain or the social compact that was forged out of the policy disputes. Society (more precisely, the federal government) would for the first time assume the responsibility for supporting basic research (mainly, but not exclusively, in the universities). Scientific manpower would be produced as a by-product of that research support. Basic research would be the engine driving the whole system. The progress of science would ensure national security, a healthy nation, and economic strength.

As noted above, the impact of the emerging Cold War and the changes in international climate were the decisive background influence that more than any other factor gave urgency to the overall debate and assured that the political system would respond. Second, science would be supported in such a way (through the judgment of scientific peers) as to ensure both the autonomy and integrity of science itself. Science was not to be wholly autonomous, but was implicitly granted wide latitude to conduct its internal affairs. Fraud and misconduct were assumed to be largely nonexistent and in any event were effectively curbed by the internal mechanisms of openness and peer competition. In return there would be, it was assumed by political elites, a flow of useful results to the government departments, such as the military, that supported the scientific advances.

The government laboratories themselves would, as recommended in the

Bush and Steelman reports, be substantially strengthened and fully articulated with the rest of the research system. The military services, with their system for supporting research all the way from basic research to field test and deployment, were the model for other government agencies.

As to the economy, several factors were critical. Government would buy complex weapons and other systems from industry rather than rely on its own arsenals or production facilities. Government would thus help new industries to emerge through "start-up" markets but would not otherwise intervene in the civilian economy. Technical manpower trained in the universities would naturally flow to industry, and the government in this sense would indirectly help maintain the U.S. edge in productivity that assured prosperity (Alic et al., 1992; Baily and Chakrabani 1987, 9). Most observers assumed that there would be no serious conflict between civilian and military needs, education and research, or basic and applied science. The needs of society and of science would be served in such a way as to protect both the independence of science and its accountability to the general public. Beyond the government's role as a buyer little explicit thought was given to what in contemporary usage is known as *technology policy.* It was assumed that the advance of basic research would bring about economic progress more or less automatically. As Harvey Brooks notes:

> The implicit message of the Bush report seemed to be that technology was essentially the application of leading-edge science and that, if the country created and sustained a first-class science establishment based primarily in the universities, new technology for national security, economic growth, job creation, and social welfare would be generated almost automatically without explicit policy attention to all the other complementary aspects of innovation. (1987, 512)

The United States thus fell in the pattern of the pluralist market-oriented system in terms of the Solingen typology advanced in the introductory chapter. The U.S. example, with such ancillary premises on the need for approximately 3 percent of GNP to be expended on R&D, was widely admired around the world, but the U.S. system was never explicitly copied by other nations, even by our close allies. The European nations, with the exception of the United Kingdom, tended to rely on a more centralized pattern of research support built around the large department of science or science and industry that was favored in the Bush report but rejected in practice (National Research Council 1982).

It may seem improbable that the postwar formula could have been so readily accepted by groups as diverse as agency research administrators, university scientists, industrialists, the news media, federal and state politicians, congressional staff, foundation officials, and others—the jumps of logic in the argument were numerous—but, the implicit and sometimes explicit claims for what science could accomplish went beyond the bounds of common sense. If held up to careful scrutiny, the ideas would have exposed unsupported assumptions, incorrect premises, and illusory hopes for harmony among potentially contending parties. But the ideas had a peculiarly U.S. virtue: they seemed to work in practice. That is, they seemed to explain satisfactorily what was actually happening. To an extraordinary degree, U.S. citizens *did* enjoy security, prosperity, advances in health, and the better life that the Bush and Steelman reports said would follow from scientific progress. The science policy compact could not be viewed as the cause of the nation's postwar economic growth or as being responsible for the West's postwar economic reconstruction. But the ideas at least appeared to explain the course of postwar events and the United States' unique position as a hegemon in the international system.

It was, in fact, logical to believe that scientific and technological superiority accounted for American economic, political, and military strength. American productivity was vastly superior to that of other nations (Baily and Chakrabani 1987, 9). U.S. firms could seemingly clobber any competitor that posed a challenge. Moreover, other nations apparently believed that the United States had found the magic formula, as J. J. Servan-Schreiber's analysis of the "technology gap" suggested (1968). Many nations sought to borrow aspects of the U.S. approaches to research support, its technology management and development practices, and its cultural openness to change (Ergas 1987, 191–245).

The unique circumstances of the postwar context helped to make the science compact possible. Scientists had helped to win the war. A grateful nation looked with deference to the views of the scientific leaders on how to organize postwar science. The scientific community was small; at the time it usually spoke with one voice or at least managed to hide internal disagreements; and it was represented by scientist-statesmen who carried broad prestige. The unusual, if only temporary, prominence enjoyed by the physicists in the initial postwar period rested on the perception that they were among the chief architects of victory. The scientists shrewdly and quietly resolved their internal differences and then publicly agreed on what resources their field needed from society. Moreover, the claims made on society were initially modest. The claims were set against the backdrop of a widely accepted need for expansion of the research and higher education system. When the scientists split publicly, as in the bitter disputes over the H-bomb development and the Oppenheimer security hearings, public veneration could quickly turn to more traditional U.S. distrust of intellectuals (Lakoff and Dupre 1962). The disputes over the H-bomb led to the development of a second weapons laboratory and to a longstanding division between the "arms control" versus the "peace through strength" wings of the scientific community (York 1987). The call for growth after the Soviet Sputnik launches in October and December 1957 helped submerge internal scientific disputes and accelerated developments already under way toward a significant expansion of the research system (Smith and Karlesky 1977; Wilson 1983, 30–42).

Growth, indeed, was an essential ingredient for the whole postwar system. The universities expanded because the nation needed more scientists and engineers for industry and government as well as to handle growing student enrollments. Unlike the Western frontier, whose closing impoverished the U.S. imagination, science was "the endless frontier." This bespoke the absence of limits, a continuous expansion of knowledge and mastery of nature. With growth benefitting all parts of the system, there were few or no losers. Conflicts were submerged in the climate of expansion. In 1965, a National Academy of Sciences panel offered the formula that funds for R&D should normally increase at a 15 percent annual rate (Committee on Science and Public Policy 1965). Growth was both a precondition for the postwar system and one of the articles of faith. Scientists could count on society's continuous support as they explored the endless frontier. Society could endorse this unwritten compact because the scientific effort as a whole would produce a continuous stream of useful results.

To help society adapt to the scientific age, it was prudent to include scientific advisers in the decision-making process. Science advisory offices were created in the individual military services, and later in the Defense Department, the Atomic Energy Commission, the National Aeronautics and Space Administration, the State Department, the Commerce Department, the Environmental Protection Agency (EPA), and other agencies (Smith 1992). A science adviser was established to serve the president as well (Herken 1992). Scientists were to advise policymakers on all matters relating to science and technology, including policies to support the growth of science. They were to participate in the inner councils of policy-making and yet remain members of the outside scientific community subject to its rules and mores.

The Breakdown of the Postwar Science-Society Pact, 1968–78

The loose cluster of unifying beliefs and attitudes on science's role described above remained intact for nearly a generation. Then the broad agreement broke down under a host of internal and external pressures. How did this happen? Why did this phase of such surprising consensus in the normally disorderly processes of U.S. democracy give way to dissensus? Partly, of course, the appearance of general agreement on both the goals and the means of science policy was an illusion. Many conflicts were masked by the rhetoric of the scientific and political leadership that tended toward the proud and self-congratulatory. The ambiguities, and even contradictions, in the cluster of ideas forming the science compact began to emerge as the research system matured and as the pace of expansion slowed. The thought that support for science could increase indefinitely at a growth rate faster than the GNP began to be recognized for what it was: a fantasy. Moreover, as the U.S. industrial performances began to falter, it became obvious that the links between basic research and economic performance were not as simple as was previously assumed (Averch 1985; Branscomb 1992, 318–21; Dickson 1984). Support for basic research declined in constant or real dollars for a decade (and even declined in actual or current dollars for several years beginning in 1967) (Smith and Karlesky 1977, 18–32). As support declined, and as the number of researchers at various stages in the pipeline continued to grow, there were more losers in the competition for funds. This put added strains on the system.

To produce highly trained manpower as a by-product of basic research support was a recipe for high social costs and a research-oriented professoriate. The federal government, in consequence, cut back sharply under the Nixon administration on fellowship and traineeship support. The aim was to slow the growth of the research system or, failing that, at least to remove subsidies for individuals to acquire skills that could be carried over to lucrative careers in industry or private medical practice.

The tumultuous events of this second phase of postwar U.S. science policy, roughly spanning the decade from 1968 to 1978, permit no simple story line. The rise of protest politics of all sorts, the consumer and generational revolt, the new gender politics, intensifying racial conflicts, environmentalism, and other matters all affected the relations between science and society. In crude terms the new tensions in science and technology policy can be inferred from the creation in 1973 of the Congressional Office of Technology Assessment. The words bespeak the notion that science (and the technologies flowing from it) needed to be "assessed" because of the numerous second- and third-order adverse consequences of technical advance. The "endless frontier" of the Bush report was no longer a benign cornucopia of good things. Science was now seen as a source of problems, environmental and otherwise. Barry Commoner's commentary typifies the extreme view of technology in an unholy alliance with industry producing a malign threat to the natural order (Commoner 1971). More serious, perhaps, than the anti-science movement of the 1970s was the shift of opinion in the political center toward a more skeptical and critical view of science and technology. Many problems, evidently, from the military conflict in Vietnam to urban decay in the central cities, were not amenable to a technological solution. The United States thus "came of age" during the crisis years of the Vietnam era and lost the utopian rationalist confidence in scientific discourse and scientism in its approach to social problems (Ezrahi 1990).

The splits between civilian and military uses of science were some of the clearest manifestations of the emerging tensions. Two seemingly minor episodes illustrate the general point. The first was the McNamara "electronic fence" that surfaced in 1968 as a technology fix to the Vietnam War (Karnow 1983, 500). In this bizarre case the political and scientific leadership in the Department of Defense apparently convinced itself for a time that a belt of electronic sensors could reveal enemy troop infiltrations and thus permit early interdiction. The second case was the adoption of the Mansfield Amendment as a means to "protect" the universities from student disturbances and campus radicals (Nichols 1971, 29–36).

The Mansfield Amendment was an action of particular symbolic importance. For it marked the definitive split between the defense and nondefense sides of the scientific community (the dispute having already been evidenced in the Oppenheimer hearings and the conflict over a second weapons lab). The adoption of the amendment also signified that the missions agencies would henceforth employ stricter standards of relevance to specific agency needs in their research support policies. The Mansfield Amendment (Military Procurement Authorization Act of 1970, Pub. L. 91-121, sec. 203), specified that

> none of the funds authorized to be appropriated by this act may be used to carry out any research project or study unless such project or study has a direct and apparent relationship to a specific military function or operation.

Many defense programs were immediately transferred to the National Science Foundation (NSF). Around this time, defense laboratories at Columbia, Cornell, Michigan, MIT, and other universities severed their campus ties. Classified research was eliminated at numerous universities. More broadly, the supposed harmony between defense needs and a strong civilian economy was clearly called into question. Defense expenditures were seen by many observers as a drain on the economy; too much of the nation's technical manpower was diverted into defense. Special programs were needed, as a corollary to this belief, to spur innovation in lagging civilian sectors (Averch 1985).

Many of the scientific community's problems came from within its own ranks. The antiscience movement that arose in the 1970s was largely the product of dissident intellectuals. The environmental movement, more moderate in tone than the antiscience movement, drew on ecologists from the universities who helped transform conservationism from romanticism to a base of intellectual respectability. Complaints on laboratory safety, discrimination by race or gender, the "old boy" network, undue corporate influence, elitism, neglected human needs, and other real or imagined ills often originated initially from students, junior faculty, and dissenters within the ranks of science and were then taken up by the politicians. The intramural quarrels drew in politicians and critics from the news media; and heightened conflict exacerbated the fissures and splits within the scientific community.

The environmental movement, which in the few years before and after Earth Day in 1970 succeeded in the passage of thirteen major regulatory statutes, reflected paradoxically both a growing distrust of science and an exaggerated faith in science left over from the early postwar era. On the one hand, critics increasingly saw science as the source of society's ills—pollution, noise, congestion, etc.—rather than as liberating mankind through scientific progress. On the other, many of the environmental laws incorporated the technological "quick fix" approach on the assumption that legislative mandates for best available technologies could force industry to develop low-cost, pollution-free alternatives to the technologies causing the problems (Jasanoff 1990; Landy, Roberts, and Thomas 1990).

On each of the key premises of the postwar consensus, then, the conventional wisdom now shifted. As regards the primacy of basic research, it was now recognized that basic research does not lead automatically to applications and that its short-run contribution to economic well-being has been exaggerated. Second, the circumscribed role of government technical bureaus as performers of applied research for their bureaus now gave way to a more active role for them in the transfer of technology to industry (Branscomb 1992, 328–36). Agencies should strive for ways to push demonstration projects and other efforts to assist the workings of the marketplace. Innovation, previously assumed to be best left to private industry, now presumably needed help from a more interventionist government. A State Technical Services Act proposed in 1965, for example, sought to create a system to assist civil sector technology development that was modeled on the agricultural extension service concept (Smith 1990, 24, 87). Where it was assumed before that science needed little regulation, science now needed extensive regulation in several senses: the by-products of and industries created by science and technology; the processes and methodologies used to measure and assess risks; and the conduct of science itself (the use of human and animal subjects, fetal tissue, laboratory safety, etc.).

Finally, the international assumptions that underlay the initial postwar phase shifted significantly. The United States was able to depart from its traditional isolationism by virtue of having the only industrial economy undamaged by war. The General Agreement in Tariffs and Trade (GATT) won acceptance from the Congress and the U.S. business community because U.S. interests would enjoy a secure international foothold and even dominance in world markets. There should be, U.S. citizens could readily conclude, free flow of ideas, goods, and services around the world, and few restrictions on the international movement of technology. However, protecting the environment, safeguarding intellectual property rights and key process technologies, and defending national security by preventing the leakage of military secrets later became much more important. Instead of the open Bretton Woods world prevailing in the 1950s and 1960s, protectionist voices emerged once again as a strong force in the United States. Competitive pressures and fears of job losses in the 1980s fueled the resentments against foreign goods and foreign direct investment.

The military alliances of the Cold War continued to exert a moderating influence on economic rivalries between the United States and its allies, however. U.S. elite opinion, though subject to stresses resulting from growing international competitive pressures, remained committed in its majority view to an internationalist outlook. The trade disputes of the 1970s and 1980s presaged the splits between the United States and its allies on economic policy that occurred once the Cold War ended. The end of the Cold War has scrambled the conceptual underpinnings for the entire U.S. research system as well as forcing a rethinking of the nation's broad strategic objectives. Whatever else a "new world order" might entail, economic objectives in virtually all Western nations have now become more important and no longer are automatically subordinate to national security considerations.

Another key element of the science pact in the United States was the presence of scientists as advisers in the inner councils of policy-making. In this respect the United States has differed from other industrialized democracies: scientists have participated broadly in "science *in* policy" issues as well as in "policy *for* science" issues (Brickman 1979; Golden 1991; Smith 1992). A science adviser and an advisory committee on science in the White House since 1957 was an important symbol of science's status (the advisory apparatus was created in the wake of the Soviet Sputniks, but it had functioned informally for several years prior to that). In January 1973 President Nixon put an end to the special position of scientists as advisers in the White House (Herken 1992; Smith 1992, 169–78). This move was the inevitable outcome of a lengthy period of tension between the "peace" scientists in the President's Science Advisory Committee (PSAC) and the president and his political advisers. The particular disputes over the supersonic transport (SST) and the Antiballistic Missile Treaty (ABM) were only the most visible manifestations of deep-seated political differences among certain PSAC members and consultants-at-large and the Nixon administration.

The move to abolish (or, more accurately, to transfer out of the White House) the advisory apparatus also was based on the belief that the PSAC scientists as outsiders did not show sufficient respect for the discipline required of close advisers to the president. Science expertise and advisory duties could also, the Nixon people felt, be better handled by less specialized staff units (in particular, by an upgraded OMB). The science advisory function was not abolished outright; the NSF developed a research staff reporting to Director H. Guyford Stever that performed some of the advisory duties of the old White House science office. Stever advised the president on occasion, mostly on energy issues and international scientific exchanges with the Soviet Union. But Nixon and his OMB director George P. Shultz felt no need for a formal scientific presence in the White House. This was a heresy to many in the scientific community who had grown accustomed to scientists sitting in the inner circles of policy-making. Clearly, however, the most important problems faced at the presidential level were not amenable to scientific solutions.

Finally, the 1968–78 decade was also marked by a shift in power away from both the executive branch and scientific advisory network toward the Congress and its subcommittees. In this respect, science policy-making simply reflected the general trend toward a more assertive congressional voice on all aspects of policy-making following the Watergate crisis. The trend markedly diminished the prospects for any elite within the scientific community to have a privileged voice in policy. Science and technology policy thereupon became demythologized, more open and contentious, and generally reflected the complex lines of cleavage in U.S. politics. As science policy became a more central feature of political debate, it reflected the social concerns that preoccupied U.S. citizens at the time and the ambiguities and confusions of politics in general.

Recent Trends

Beginning in the middle of the Carter term, and accelerating under Presidents Reagan and Bush, the nation struggled to refashion the remnants of the postwar science pact into a new approach to the relations of science and society. The aim was to recreate a solid framework of research support and restore an important role for science in national defense and economic growth. Deference to scientists as such could not be reestablished; the conflicts of the previous decade ruled out any simple return to the past. The unique circumstances that had permitted a relatively small group of executive branch officials and an elite corps of physical scientists largely to set the postwar science policy agenda no longer existed. The scientific community, it seemed likely, would never enjoy the same broad deference from the rest of society.

Yet the past evidently was prologue. For policymakers seemed to retrace steps already taken. After a decade of crisis in government-science relations the traditional relationships were partially restored. This was evidenced in several ways: basic research funding was substantially increased and applied research and development efforts were cut back; defense expenditures increased as a percentage of total federal R&D (up from 50 percent in 1978 to approximately two-thirds by 1988) (Teich 1991); and the Reagan administration, at least initially, sought to back off from intervention in the economy in favor of reliance on deregulation and market incentives (Dickson 1984; Smith 1990, 128–35). The renewed attention to basic research, after a decade of slow growth in federal funding, occurred despite fiscal stringency. The assumption seemed to be that if basic research could not be shown clearly to foster short-term economic performance, at least it might contribute to long-run growth. More important was the link between the failure of Nixon-era détente and the renewed external threat and support for increased R&D spending. The Reagan administration focus on defense, and the perceived failure of the détente pursued under Nixon and under Carter until the Afghanistan invasion, created a climate receptive to increased R&D spending.

Moreover, the Reagan administration in its second term made doubling the NSF budget over a five-year period a centerpiece of its economic competitiveness strategy. The plan was rebuffed by Congress, however, which scaled back the request. Industry, stimulated in part by the growth in defense spending, increased its own R&D outlays (though in some industries R&D spending by the private sector decreased again after 1984) (Brooks 1986, 119–67; National Science Board 1988).

However, the pieces of the old formula did not altogether fit with the new policy thrusts. Science policy resembled a jigsaw puzzle with the pieces not quite matching. A close look at the 1980s reveals numerous incongruities and departures from the past. The mission agencies—the Department of Defense (DOD), in particular—no longer fully embraced the earlier view that they should support broad mission-oriented basic research as an "overhead" on their narrower technical activities. The Mansfield Amendment thinking, in effect, carried over into practice, even though it formally disappeared from the statute books. When the DOD later in the Reagan presidency sought to rebuild strong university ties, it bungled the effort by trying to exercise more control over program content than the DOD had in the earlier period. The paradoxical result was that when DOD support was a larger share of total university research support, the control exercised was less. The defense agencies earlier had permitted and even encouraged university autonomy. Friction persisted between universities and the DOD throughout the 1980s, although tensions were eased by the creation of the DOD–University Roundtable in 1983.

A part of the problem was the clash between national security officials and the research community over secrecy and classification issues. National

Security officials in the first Reagan term in one dramatic case forced researchers to withdraw papers from open discussion at a mathematics conference in San Diego. The debate over the openness of scholarly communications, and the right of defense officials to censor papers or restrict the access of foreign scientists to U.S. labs, was part of the larger issue of the administration of export control laws (National Research Council 1987).

A new element in the United States' approach to science policy under Presidents Reagan and Bush has been the emergence of ideologically oriented controls on the conduct of research stemming from the right wing of the Republican party. The abortion issue, and its complex politics, has been a critical factor. Federal support for research using fetal tissue, in vitro fertilization, and the abortion pill RU-486, for example, have also been proscribed. Debate arose in the first two years of the Bush administration over the issue of whether high-level scientific appointees must be subject to an abortion "litmus test." An eighteen-month delay in filling the post of National Institute of Health (NIH) director was the result.

More broadly, beyond the abortion issue, conflict arose over recombinant DNA research, research involving human subjects, animal rights in the laboratory, and scientific fraud and misconduct (LaFollette 1992). The first two concerns—recombinant DNA research and the consent of human subjects—have been largely resolved through administrative action and without recourse to legislation. Animal rights and scientific fraud and misconduct have proved to be more refractory issues. Animal rights activists have been successful in bringing about legislation mandating clean laboratory conditions, but this has not satisfied the more extreme wing of the movement. The militant People for the Ethical Treatment of Animals (PETA) has been implicated in a number of violent incidents directed against the laboratories of scientists who conduct research using animals.

Since 1985 the NSF and NIH have required universities to develop policies and procedures for handling scientific misconduct cases and complaints. The original slow response of the scientific community to reports of fraud and misconduct concerned the Congress that formal requirements were necessary, and the much-publicized case involving congressional testimony by Nobel laureate David Baltimore before a subcommittee chaired by Congressman John Dingell merely reinforced the sentiment that scientists were foot-dragging and inadequate in their response to misconduct concerns.

The women's movement in the early 1990s forced a modification in NIH policies on the conduct of clinical trials in biomedical research. Researchers were required to include women as subjects in clinical trials unless the nature of the research project would clearly make their inclusion impracticable. A new division was also established to focus on women's diseases.

To some members of the scientific community, such government actions suggest the potential for political control of science. Fears by pre–World War II scientists who opposed federal support for science because of potential federal controls on scientific autonomy seem to have been validated. To others the steps seemed more like extensions of traditional U.S. attitudes favoring narrowly applied rather than basic research and utilitarian demands for more relevance in what scientists do. Either way, the message has seemed clear: the postwar formula that left the scientific community largely free to set its own agenda while receiving large amounts of research support is in a continuing process of amendment and redefinition. Federal support for research will now carry with it greater obligations both as to how research is done and as to tangible benefits for society.

Congress has also exerted greater influence in the allocation of funds for university research facilities. Such actions are decried as "pork barrel," and as a potential threat to peer review in the allocation of research funds generally. Congress nevertheless continues to earmark funds for particular universities and regions. The "have-not" universities defend the practice as the only way to achieve greater geographic equity in overall R&D funding. To some critics the politicization of science that earmarking represents could jeopardize the critical pillar of the whole postwar science relationship: peer or merit review of proposals. This process has helped assure that resources are efficiently allocated to the highest-quality research performers. If the more political process that exists with large-scale projects and grants for facilities were to spread to resource allocation decisions generally, critics fear a permanent loss of scientific autonomy. Large-scale science and technology programs have, of course, always tended to attract congressional interest and to reflect political pressures (Cohen and Noll 1991).

Vannevar Bush, in *Science—The Endless Frontier,* had urged that science should move from "the wings to center stage" of public concern and attention. Science has clearly moved to center stage in recent years, but in doing so it has ceased to be a cloistered community removed from the rough and tumble of U.S. politics. Science has become part of the clamorous political mainstream and has increasingly reflected its contradictions and confusions.

Among the most critical science and technology policy issues facing the nation are the issues of (1) competitiveness in the new global economy and (2) the full implications of a shrinking defense R&D base in the face of the end of the Cold War. To begin with the competitiveness issue, the Reagan administration began with the conventional view that innovation was a matter for the private marketplace, not of government policy. Then it moved gradually (or was pushed by Congress) toward policies to stimulate private R&D investment and innovation. These policies included the creation of the manufacturing consortium Sematech, passage of the Federal Technology Transfer Act in

1986, the relaxation of antitrust standards impeding R&D consortia in 1984, the creation of the NSF engineering research centers, research support by Defense Advanced Research Project Agency (DARPA) for superconductivity and high definition television (HD-TV), and trade legislation at the end of the Reagan presidency that created the Advanced Technology Program in the Department of Commerce and broadened the responsibilities of the National Bureau of Standards (renamed the National Institute of Standards and Technology), and other initiatives. The support for Sematech, HD-TV, and other targeted research efforts by the Departments of Defense and Commerce evidenced the Reagan administration's more pragmatic side.

The Bush administration in September 1990 took the step of announcing a technology policy stressing the importance of government support for "generic precompetitive technologies," and an ambitious program of building an information infrastructure for the nation (Branscomb 1992; Executive Office of the President 1990; Graham 1992). The Bush action was the first time that a president and an administration had explicitly announced the need for a policy toward civilian technology development. While the policy was cautious and stopped well short of broad government intervention into the economy, the move strengthened those within the government who had favored "industrial policy." The Commerce Department, for example, dating back to the 1920s under the Herbert Hoover secretaryship, had sought such a role (Dupree 1957). In the 1930s, the National Bureau of Standards attempted to win congressional approval for a program of research to be performed on behalf of U.S. industry. The proponents evidently had in mind something akin to what would later be called mission-oriented basic research. But, as Carroll W. Pursell, Jr., notes, "the failure to define 'basic' adequately left the measure open, on one side, to charges of ivory-tower dreaming and, on the other, to fears that manufacturing processes would be patented and licensed by the Department of Commerce" (1968, 163; 1979 162–74).

After World War II, the Commerce Department and the National Bureau of Standards (NBS) had an ambitious plan for federal funding of civilian industrial research. NBS had built the first working modern computer, the SEAC, but the plan ran into opposition, partly as a result of the scandal over the AD-X2 battery additive that had riled the business community (Flamm 1988; Lawrence 1962; Pennick et al. 1965). Under Secretary Robert Mossbacher early in the Bush administration the Commerce Department had sought a major role in pushing for the rapid commercialization of HD-TV, but had been rebuffed (Beltz 1991).

The future of the defense technology base was an especially complex issue for the nation. As the threat from the Soviet bloc began to erode and then virtually disappear, friends as well as critics of the military began to wonder what level of defense R&D effort was still needed. The rationale for the large national security technology effort came under serious scrutiny. For example, if drastic cuts were to occur in defense spending, should the nation stress the readiness of the smaller forces remaining? Or should the nation, conversely, pursue modernization and new technologies even at some cost in force readiness? If modernization were chosen, the issue was then whether R&D should be emphasized to provide the base for future systems developments or whether the nation should push ahead with procurement and deployment of new systems themselves. To what extent could the nation emphasize an R&D strategy that would "stockpile" ideas for possible later development or was it necessary to produce something to keep the defense technology base (and jobs) in a healthy condition?

In any case, the resources in the defense budget for the technology base would be drastically shrunk from present levels. Such cuts would have ripple effects through the entire research system from the universities to industry to the national laboratories. The three sectors—industry, universities, national labs—might end up competing with each other for shares of the smaller pie. In the past the three sectors had coexisted comfortably along the lines of a clear division of labor among basic research, applied research, and procurement. A strong civilian economy with a swift pace of innovation now seemed to provide the best technical base for the future military establishment. The U.S. military in the past had functioned something like the command economy of the former Soviet Union: a "dual economy" of separate civilian and military components had existed. In the early postwar period the military sector was technologically more advanced. Increasingly, however, the pace of civilian technology development outstripped developments with the procurement-dominated military sector. As Kenneth Flamm and Thomas McNaugher conclude: "political incentives and an encrusted organizational structure [have] had unfortunate consequences for the quality of U.S. military R&D generally and more specifically for its usefulness in generating commercial returns"; and "if the nation needs a stable effort to develop its technology base, the defense budget is not the place to lodge the effort" (1990, 141, 147). A separate military and civilian technology base in the United States, in short, was no longer realistic or desirable (Alic et al. 1992; Branscomb 1992).

Moreover, the whole research system seemed likely to shrink in the force of defense cuts with perhaps unpredictable consequences for the health of the underlying scientific base. Edward David, Jr., former science adviser to President Nixon and retired chief of research for Exxon, believes that the Cold War's end will bring a 25 to 30 percent drop in federal funding for R&D (Marshall 1992, 882).

U.S. industry, in the face of the new developments, has been going through a period of soul-searching. Industrialists have wondered whether the problems with U.S. competitiveness are the result of macroeconomic policy failures, scientific failures, failures in the "downstream" aspects of manufacturing, labor-management failures, or some combination of these various factors. Most seem to have concluded that the emphasis should be on quality assurance and process improvements on the shop floor rather than on macropolicy initiatives, though protectionism has gained ground and appeals for clearer policy leadership from the U.S. government have grown more insistent. Industry has been deeply engaged in the reorganization of its own technical activities; in many cases, companies have broken up their central research laboratories and distributed the R&D functions to the business units. The 1992 presidential election brought considerable attention to the issues of technology policy, and initially it seemed likely that President Clinton would pursue a more aggressive policy than his Republican predecessor Bush administration in the area of civilian technology. But as of June 1993 the priorities of the new administration were weighted toward short-run economic recovery and no concrete long-term technology strategy had yet emerged despite a great many rhetorical pronouncements.

The U.S. scene has been one of contrasts, unfolding changes, and lively controversies in the 1990s. Most U.S. citizens believe that science is important to a wide variety of national goals, and that a healthy scientific community is a national priority. But they no longer seem willing to accept the formula that science should have a special privileged status in the policy process or that the scientific community should necessarily define its own priorities. Science is seen as one voice among many that must be heard in deciding important issues. Society (or to speak less anthropomorphically, its elected representatives) apparently feels entitled to take a more active role in seeing that public resources advance tangible social objectives. The exact roles of the different research institutions remain unsettled in this time of rapid change. The case has to be made anew for the best organization of science affairs in the larger national interest. Both the National Science Foundation and the National Institutes of Health were engaged in strategic planning reviews at the end of 1992 that apparently were at least in part responses to congressional pressures for a reorientation of the national research effort toward more utilitarian goals.

Possible Future Developments

The U.S. research system thus stands in a position of both promise and threat at the time of this writing in June 1993. It is the best of times for science and the worst. By measures such as receipt of Nobel prizes and publication rates, U.S. science evidently remains at the forefront in many disciplines (OECD 1988, tables 5, 6.1). Basic science in the universities has continued to flourish, but the supply of talented researchers with good ideas has apparently outstripped society's ability to provide adequate support for them all. With research budgets growing but only slowly, and with science becoming a more expensive and capital-intensive activity, constraint has become a fact of life for the scientific community in the United States. In this respect the U.S. pattern follows the austere conditions that have prevailed in the United Kingdom for nearly a decade and the fiscal constraints that have also affected other European scientists in recent years.

The changes in the U.S. research system will be felt in virtually all sectors. In the universities the assumptions that have guided administrators for most of the entire postwar period are being reexamined. The system has to adapt in some fashion: perhaps by concentrating research in a smaller number of elite universities as was the pattern before World War II. Or productivity gain must be achieved in some other fashion; e.g., through centers that share facilities, changing the mix of students to technicians in research, more specialization of effort among institutions, etc. The tradition of Max Planck institutes organized apart from universities is a possible future development, but the elitism that this might imply would be bound to generate intense controversy. Universities may be forced to "down-size" just as industry and government have undergone significant changes and restructurings.

The linkage of teaching and research has traditionally been considered a strength of the U.S. university research system, as with the wide dispersion of research capacities thereby implied. To adopt a system of research institutes separate from the universities or to create fewer centers of scientific excellence in a smaller number of universities would be significant departures from the past. Yet change may be needed to avoid having the whole system gradually lose momentum. The loss of defense research seems bound to pose acute problems for a number of universities and to have special impact for various fields (e.g., 40 percent of research funding for artificial intelligence reportedly comes from the DOD). The universities have also come under intense scrutiny from government auditors over abuses in the indirect cost reimbursements under government research grants and contracts.

Yet the universities, for all of their problems, may still be better off than some other parts of the research system. The other parts of the system are perhaps facing even greater challenges. The government laboratories, almost 800 entities spending nearly $20 billion a year, suffer from a number of problems. The shrinking defense budget almost certainly means that some of them will be shut down, reorganized, merged, or scaled back in the size and scope of their activities. At the same time they have been charged with a broader role in technology transfer to industry. Although "conversion" to civilian needs is talked about as a possible answer to excess defense capacity, the nation has never succeeded in converting military laboratories or industry to civilian uses. In the past, when defense contracts have been canceled, plants have simply closed down and laid off scientists and engineers.

Even more significant problems may be faced by industry. How does the nation modernize its industrial base to be competitive in the world economy? The problems include a technologically-backward supplier base, a narrow concentration of research capacity among relatively few large firms, lack of cooperation among firms on standards and quality assurance procedures, barriers between the research, production, and marketing departments in many firms, inadequate investment in long-range research on promising new technologies and processes, and excessively long product cycles. U.S. industry does some kinds of research well, but often fails to invest enough in exploiting new scientific discoveries in a timely fashion. In particular, U.S. firms seem to have trouble in the product realization process (Dertouzos, Lester, and Solow 1989, 154, 157).

The United States needs to upgrade, and perhaps to revise radically, its industrial research operations. One possible line of development is a shift of a larger share of the nation's total R&D effort to industry; i.e., U.S. companies might become more like Japanese companies in large-scale support of advanced training, basic and applied research, and product development and testing. Many U.S. companies, however, have been cutting back on basic research, dismantling their large central labs as expensive overhead, and focusing on short-run productivity gains. If U.S. industrial capacity were to erode further, political pressures for "industrial policy" such as arose over the recessionary 1980–82 period, could intensify. An apparently strong economic recovery in 1982 helped to remove the impetus for major policy steps. But the pressures for more government intervention could reemerge if economic recovery were to be robust in 1993.

A potentially promising line of development is new partnerships between companies, between universities and companies, and between governmental labs and companies. Experiments are underway in all of these areas. Universities could be drawn even more closely into the world of business and the early stages of product development. There are incentives that could move events in this direction, leading the universities into a more stratified system with clear lines between those with primarily basic research functions and those with applied research functions. Most of the large universities continue, however, to tilt toward a pure research model, and many shy away from collaboration with industry. Industry could probably gain from funding partnerships with universities on generic applied technologies. Industry seems to have virtually abandoned a basic research role, which could create both problems and opportunities for the universities. A company's research department was the most natural point of contact with the academic world. With such labs transferred to the business units or abolished outright, the natural points of contact with the universities are limited. At the same time, industry may find it useful to contract with universities for some of its long-range research effort. Product development responsibilities, however, will likely remain firmly in company hands.

For all the pressures toward change, the pressures toward system maintenance in the United States are not to be discounted. A great deal of experimentation, trial and error adaptation, and shifts of emphasis have always taken place, but the core institutional actors—universities, government labs, industrial research units, not-for-profit research institutes—have in the past continued to have distinct functions and to find an "ecological niche" in the overall system.

If the drive toward market mechanisms continues at the same rapid pace in the nations of the former Soviet Union, the international movement of ideas, technologies, scientific personnel, joint ventures, and other pooled technical effort might transform the research systems of all nations. The new megafirm, with far-flung technical operations, could dwarf the multinational corporations of the 1970s and 1980s. With the dwindling East-West national security threat, export controls should be eased. The rise of new powers (e.g., a revived Iran, a nuclear North Korea, an arms race in the Pacific) makes it likely that the industrial nations will try to slow the spread of some technologies so that export controls will not vanish altogether.

New flows of technologies could spark recovery from the worldwide recession and even a surge of prosperity, the spread of scientific habits of thought, and a new international division of labor. Scientific institutions, which have always had an international outlook, could become even more internationalist in staffing, operations, and outlook. Technologies will never flow completely freely around the world, embedded as they are in proprietary interests. But if a secure regime of intellectual property rights can be established, U.S. business will increasingly embrace a global focus. In these developments the research system of all nations will face stresses and be forced to adapt to new circumstances.

The research systems of the industrial nations will be both the objects and the agents of change. The U.S. research system, marked by openness and fluidity, and with its noisy process of reaching compromises and mutual accommodations among scientific and political actors, may yet prove to be more adaptable than centrally planned systems.

BIBLIOGRAPHY

Alic, John, Lewis M. Branscomb, Harvey Brooks, Ashton Carter, and Gerald Epstein. 1992. *Beyond Spinoff: Military and Commercial Technologies in a Changing World.* Cambridge, MA: Harvard Business School Press.

Averch, Harvey. A. 1985. *A Strategic Analysis of Science and Technology Policy.* Baltimore: Johns Hopkins Press.

Baily, Martin Neil, and Alok K. Chakrabani. 1987. *Innovation and the Productivity Crisis.* Washington, DC: Brookings Institution.

Beltz, Cynthia A. 1991. *High-Tech Maneuvers: Industrial Policy Lessons of HD-TV.* Washington, DC: AEI Press.

Branscomb, Lewis M. 1992. "America's Emerging Technology Policy." *Minerva* 30, no. 2 (Autumn): 317–36.

Brickman, Ronald. 1979. "Comparative Approaches to R&D Policy Coordination." *Policy Sciences* 11 (August): 73–91.

Brooks, Harvey. 1987. "What Is the National Agenda for Science, and How Did It Come About?" *American Scientist* 75 (September-October).

Brooks, Harvey. 1986. "National Science Policy and Technological Innovations." In *The Positive Sum Strategy: Harnessing Technology for Economic Growth,* ed. Ralph Landau and Nathan Rosenberg, Washington, DC: National Academy Press.

Bruce, Robert V. 1987. *The Launching of Modern American Science, 1846–76.* New York: Knopf.

Bush, Vannevar. [1945] 1990. *Science—The Endless Frontier.* Washington, DC: National Science Foundation.

Cohen, Linda R. and Roger G. Noll. 1991. *The Technology Pork Barrell.* Washington, DC: Brookings Institution.

Committee on Science and Public Policy, National Academy of Sciences. 1965. *Basic Research and National Goals.* Report to the House Committee on Science and Astronautics. Washington, DC: National Academy of Sciences.

Commoner, Barry. 1971. *The Closing Circle: Nature, Man, and Technology.* New York: Knopf.

Dertouzos, Michael L., Richard K. Lester, and Robert M. Solow. 1989. *Made in America: Regaining the Productive Edge.* Cambridge, MA: MIT Press.

Dickson, David. 1984. *The New Politics of Science.* New York: Pantheon.

Dupree, A. Hunter. 1957. *Science in the Federal Government: A History of Policies and Activities to 1940.* Cambridge, MA: Harvard University Press.

Dupre, J. Stefan, and Sanford A. Lakoff. 1962. *Science and the Nation.* Englewood Cliffs, NJ: Prentice Hall.

Ergas, Henry. 1987. "Does Technology Policy Matter?" In *Technology and Global Industry,* ed. Bruce R. Guile and Harvey Brooks. Washington, DC: National Academy Press.

Executive Office of the President, Office of Science and Technology Policy. 1990. *U.S. Technology Policy.* Washington, DC.

Ezrahi, Yaron. 1990. *The Descent of Icarus.* Cambridge, MA: Harvard University Press.

Flamm, Kenneth. 1987. *Targeting the Computer.* Washington, DC: Brookings Institution.

Flamm, Kenneth. 1988. *Creating the Computer.* Washington, DC: Brookings Institution.

Flamm, Kenneth, and Thomas McNaugher. 1990. "Rationalizing Technology Investments." In *Restructuring American Foreign Policy,* ed. John Steinbrunner. Washington, DC: Brookings Institution.

Geiger, Roger L. 1986. *To Advance Knowledge.* Oxford: Oxford University Press.

Golden, William T., ed. 1991. *Worldwide Science and Technology Advice to the Highest Levels of Government.* New York: Pergamon.

Golden, William T., ed. 1980. *Science Advice to the President.* New York: Pergamon.

Graham, Margaret B. W. 1985. "Corporate Research and Development: The Latest Transformation." *Technology in Society* 7:179–95.

Graham, Otis, Jr. 1992. *Losing Time: The Debate on Industrial Policy.* Oxford University Press.

Herken, Greg. 1992. *Cardinal Choices.* Oxford: Oxford University Press.

Hounshell, David A., and John Kenly Smith, Jr. 1988. *Science and Corporate Strategy: DuPont R&D, 1902–1980.* Cambridge: Cambridge University Press.

Jasanoff, Shelia. 1990. *The Fifth Branch.* Cambridge, MA: Harvard University Press.

Karnow, Stanley. 1983. *Vietnam: The History.* New York: Viking Press.

Kevles, Daniel J. 1990. Introduction to *Science—The Endless Frontier,* by Vannevar Bush. Washington, DC: National Science Foundation.

Kevles, Daniel J. 1977. "The National Science Foundation and the Debate Over Postwar Research Policy, 1942–45." *Isis* 68.

Kuznick, Peter J. 1987. *Beyond the Laboratory.* Chicago: University of Chicago Press.

LaFollette, Marcel. 1992. *Stealing Into Print: Fraud, Plagiarism, and Misconduct in Scientific Publishing.* Berkeley and Los Angeles: University of California Press.

Lakoff, Sanford A., ed. 1966. *Knowledge and Power.* New York: Free Press.

Lakoff, Sanford A. and J. Stefan Dupre. 1962. *Science and the Nation.* Englewood Cliffs, NJ: Prentice-Hall.

Landy, Mark K., Marc J. Roberts, and Stephen R. Thomas. 1990. *The Environmental Protection Agency: Asking the Wrong Questions.* Oxford: Oxford University Press.

Lawrence, Samuel A. 1962. *The Battery Additive Controversy,* Cases in Public Administration. University of Alabama Press.

Marshall, Eliot. 1992. "NSF: Being Blown Off Course?" *Science* 258, no. 5084 (6 November): 880–82.

National Research Council. 1987. *Balancing the National Interest: U.S. National Security Expert Controls and Global Economic Competition.* National Academy Press.

National Research Council. 1982. "Research in the U.S. and Europe." In *Five Year Outlook on Science and Technology.* San Francisco: Freeman and Co.

National Science Board. 1988. *Science Indicators, 1987.* Washington, DC.

Nichols, Rodney W. 1971. "Mission-Oriented R&D." *Science* 172 (2 April): 29–36.

OECD. 1988. *Science and Technology Outlook, 1988.* Paris.

Pennick, James L., Jr. et al., eds. 1965. *The Politics of American Science, 1939 to the Present.* Chicago: Rand NcNally.

Price, Don K. 1965. *The Scientific Estate.* Cambridge, MA: Belknap Press, Harvard University Press.

Price, Don K. 1954. *Government and Science.* New York: New York University Press.

Pursell, Carroll W., Jr. 1979. "Government and Technology in the Great Depression." *Technology and Culture* 20 (January): 162–74.

Pursell, Carroll W., Jr. 1968. "Legislation and the National Bureau of Standards." *Technology and Culture* 9 (April): 163.

Reich, Leonard S. 1985. *The Making of American Industrial Research: Science and Business at G.E. and Bell 1876–1926*. Cambridge: Cambridge University Press.

Sapolsky, Harvey M. 1990. *Science and the Navy*. Princeton, NJ: Princeton University Press.

Servan-Schreiber, J. R. 1968. *The American Challenge*. New York: Atheneum.

Skinner, Wickham. 1985. "The Taming of the Lions: How Manufacturing Leadership Evolved, 1780–1984." In *The Uneasy Alliance: Managing the Productivity-Technology Dilemma,* ed. Kim B. Clark, Robert H. Hayes, and Christopher Lorenz. Cambridge, MA: Harvard Business School Press.

Smith, Bruce L. R. 1992. *The Advisers: Scientists in the Policy Process*. Washington, DC: Brookings Institution.

Smith, Bruce L. R. 1990. *American Science Policy Since World War II*. Washington, DC: Brookings Institution.

Smith, Bruce L. R., and Joseph J. Karlesky. 1977. *The State of Academic Science*. New Rochelle, NY: Change Magazine Press.

Steelman Report. 1947. *Science and Public Policy: Administration for Research*. 3 vols. Washington, DC: GPO.

Teich, Albert et al. 1991. *AAAS Research XIII Research and Development FY 1989*. Washington, DC: AAAS.

Wasserman, Neil H. 1985. *From Invention to Innovation: Long-Distance Telephone Transmission at the Turn of the Century*. Baltimore: Johns Hopkins University Press.

Wilson, John T. 1983. *Academic Science, Higher Education, and the Federal Government, 1950–1983*. Chicago: University of Chicago Press.

Wise, George. 1985. *Willis R. Whitney, General Electric, and the Origins of U.S. Industrial Research*. New York: Columbia University Press.

Wolfle, Dael. 1972. *The Home of Science: The Role of the University*. New York: McGraw Hill.

York, Herbert. 1987. *Making Weapons, Talking Peace*. New York: Basic Books.

France: Science within the State

Frank R. Baumgartner and David Wilsford

Etel Solingen argues in the introduction to this volume that the internal nature of a regime and the international context within which it operates combine to produce a national style of relations between scientists and the state. The presence of such a style could be neither more clear nor more important than in France. The distinctive international posture of France since World War II, its desire to ensure continued international *grandeur,* its involvement in many international conflicts, and its status as a former colonial power make science policy of great importance to the nation's elite. Science is seen as a means through which France may maintain its international stature. Similarly, the domestic structures of French government, administration, and society conspire to leave their imprint on the relations between scientists and the state. The state, through its large centralized research organizations, dominates science policy in France. Indeed, science and the state are so closely intertwined, their relations so symbiotic, that we can rightfully present a model of fusion. Science policy-making and most scientific research activities in France are in one way or another almost entirely encompassed by the state. Science in France is led by the state.[1]

Scientific innovation in France has been prized by state policymakers in virtually every policy domain in which science is relevant. Both economic

1. There are numerous good histories of this traditional pattern of state-science relations in France, as well as of the peaks and valleys that are part of the pattern. The importance of science and engineering to the rise and consolidation of the *ancien régime* is treated in Caullery 1948. The beginnings of "modern" science and its integration into the modern French state with the Revolution and the Napoleonic period is treated in Dhombres 1989 and Rashad 1988. The consolidation of rational science and its role in the Third Republic recounted in Paul 1985. The advent of modern physics in the early 20th century is described by Pestre (1984). The interwar period and the rise of a modern French technocracy is treated in Brun 1985. See also Gillispie 1980 for the latter part of the eighteenth century prior to the Revolution, Fox and Weisz 1980 for the nineteenth century generally, and Janicaud 1987 for an overview of the twentieth century. Gilpin 1968 presents the classic account of the postwar years; Scheinman 1965 gives the best overview in English of nuclear power policy in the immediate postwar period.

development and military security reasons are used to justify this interest, and these justifications have changed little over the centuries. Not only did Louis XIV organize, subsidize, and regulate the saltpeter and gunpowder industries to further his consolidation of the French realm, but he did so in order to further his numerous foreign military ventures. Likewise, science in its many basic and applied forms is central to the French state's furtherance of its perception of its interests today, from the search for economic competitiveness of French industry in the interdependent world economy to the use of French armaments and associated military technology and equipment in France's own military engagements (from Tchad to the Persian Gulf) and the marketing of military technology to a number of foreign countries, such as Argentina and Iraq.

For economic competitiveness in industry, for the marketing of French arms and associated military technology, and for maintaining France's position in the world, state elites see scientific progress as a key national asset that must be maintained. Throughout the history of these relations, France's standing among the world's nations has been key, for France has always been sensitive about its role and place in the international community. Further, relations have been structured by rigid hierarchies designed to support planning and top-down coordination. In the French view, planning should be comprehensive (that is, covering the entire national territory), and designs should be standardized (bridge specifications, canal locks, railroad beds in earlier periods, nuclear power plants more recently) (Smith 1990). These two features have characterized science policy and the development of new industries throughout modern French history. Unlike U.S. citizens, who tend to consider most government planning exercises by definition to be ill-conceived and probably doomed, the French, like the Japanese, have historically had faith that government planning, while not perfect, should be attempted. The history of state-science relations in France shows that science and scientific planning have always been central to the state's efforts to achieve its goals.

The history of French science also shows the dominance of engineers and applied science over basic research, much as in Japan. Canals, roads, bridges, railroads, and other kinds of infrastructure have always assumed greater importance than theoretical progress alone. Railroads, for example, were a focus of centralized planning as early as the 1830s, and they remain central to French industrial policy today. French planners soon expect to link all the main cities of France into a network of high-speed (*trains à grande vitesse* or TGV) bullet train lines; they also expect to link this TGV network with the major European cities of Britain, Germany, Italy, and Benelux, as well as to Scandinavia.

Aerospace, arms, and nuclear energy are also modern concerns of French planners. In aerospace, both military and civilian aircraft have been emphasized, beginning with the Caravelle of the 1950s (the first jet to compete with U.S. producers), the Concorde of the 1970s (the only supersonic airplane developed and brought to market in the West), and the Airbus of the 1980s and 1990s (the only non-U.S. jumbo carrier to compete with Boeing). Military aircraft are, if anything, even more important to the French. The Mirage and Rafale have long been acknowledged to be among the best in the world, and remain important export items. Moreover, the company that makes them, Dassault, enjoyed such a symbiotic relationship with the state that the company's nationalization in the mid-1980s only marginally modified this relationship (Kolodziej 1987). Even as a private firm, Dassault commonly followed the lead of the state in its R&D projects, thus illustrating another notable feature of French state-science relations: even private sector R&D is led by the state.

Finally, the Ariane rocket, developed by a public-private consortium, has proved to be a technological and commercial success, especially in the period after the U.S. spaceshuttle Challenger disaster, and other mishaps, placed the U.S. space program seriously behind schedule. Many commercial clients who had planned payloads on U.S. rockets or on the spaceshuttle turned to the Ariane instead. In all these cases, the concern of the French state has been more with the development of a functioning industry capable of demonstrating French technological prowess on a grand scale rather than only with basic research.

French technocratic elites see state leadership as essential in several ways. First, national prestige and military independence depend on scientific innovation, and the French state is willing to invest heavily in projects that might serve these goals. Second, public works contracts for major infrastructure projects such as railroads, airplanes, power generation, and telephones are used to support French industry. Third, and definitely last, these investments, leaders hope, will sometimes make French products competitive on the world market. Traditionally, national prestige has come first, with marketplace success being a kind of bonus. The most successful cases serve all three of these goals, but many serve only the first two. While this mentality comes under increasing economic pressure each year, it remains a distinctive characteristic of French science policy in the 1990s.

French scientists and the bureaucrats who employ them are imbued with the notion that France has an interest separate from any particular interest of a French person, company, or group. As for Rousseau, the general will is different from, and of course superior to, any particular wills, and the responsibility of the ruling elite is to define the national interest and to use all the powers of the state to pursue it. In the face of this Cartesian positivism, critics may be dismissed as protectors of some parochial interest of the diverse parts that make up France. Rationalism and technique allow man to dominate nature, according to this view, and the state will be at the forefront of assuring that the needs of the nation are met.

The administrators and engineers at the top of the French state are specialists in the techniques of decisionmaking and of governance, not in the various substantive domains of government policy. That is, they are trained to believe that the engineering and problem-solving skills that they have learned in the nation's most prestigious schools can be used in virtually any domain. The notion of *polyvalence,* or being useful for many purposes, has a long history even in the relatively technical areas of French administration. For example, when a commission was appointed in the nineteenth century to lay out the grand lines of the French national railroad, one of France's most celebrated engineers, Louis Navier, was called to be its head. This commission went on to plan out the distinctive pattern of grand lines going out of Paris, with secondary lines feeding out to virtually every part of France. This design came to be known as the Legrand Star, and it was considered a great success. But "neither Navier nor any other engineer appointed to the Legrand Star commission in September 1832 possessed any practical experience of railroad building" (Smith 1990, 668). Similarly today, high civil servants move freely from one ministry or state agency to another, rarely remaining in a single policy sector for their entire careers.

As Victor Legrand exclaimed while setting up the engineering team that would plan the Legrand Star, "What a fine role for the State if it can take charge and plan the main lines . . . and by this means of rapid, long-distance communications bring about the full integration of our fine country" (Willemain 1862; cited in Smith 1990, 668). These plans would of course be implemented by private companies during the nineteenth century, but not without the firm hand of the state guiding their investments. Pierre Laroque, formerly Director-General of Social Security, the founding father of the postwar French social security system, and one of the great postwar civil servants who rebuilt France, gave a modern statement typical of this perspective: "The state is the juridical personification of the nation. The state *is* the nation."[2]

A recent book by two young high civil servants provides another restatement of this common viewpoint (Olivennes and Baverez 1989). They discuss the inadequacies of public bureaucracy and its dealings with the public, but their feelings are tempered by their sense of mission and security reflected in the subtitle of their book, *L'Etat, c'est nous.* For senior civil servants, even at the beginnings of their careers, this is a common point of view, reflecting their sense of responsibility for the future of the country, their view that they should

2. Laroque made the statement in a public address at an international symposium on social security systems sponsored by the French government (Paris, 13 June 1990).

decide what is the nation's best interest, and their assurance that they know best how to enact policies to achieve the nation's goals.

Because of the long-lasting, strongly-held, and widely-shared desire among French elites to use science and technology to enhance the country's international stature, science policy in France tends to be state-dominated, centralized, and focused on a small number of prestigious projects. Basic research and fundamental science are given less attention than high-visibility and large-scale science and engineering projects that have more immediate applicability. Science at the service of the state has had tremendous success in many areas, including France's military and civilian nuclear power programs, its telecommunication system, the Concorde and Airbus aircraft, high-speed trains, and the Ariane space program. On the other hand, France lags behind other countries in those areas of basic research that have not been defined as national priorities (Cohen 1992; Jobert and Muller 1987; Salomon 1986). In all this, France's scientific community has been greatly influenced by the country's position in the international system and by the internal organization of political structures. We address each of these factors in turn.

Grandeur and the Uses of French Science

In the preface to his comprehensive overview of the relations between the state and science under the *ancien régime,* Charles Gillispie notes that "much of science has in general little or nothing to do with government, and . . . much of government has little or nothing to do with science, but . . . there are intersections" (1980, ix). Indeed there are. The French state has been active for centuries in the promotion of science because it has seen science as a means toward achieving a variety of domestic and international goals. Both economic development and foreign military goals have justified the state's interest in science. Indeed, Elie Cohen has argued that almost every major industrial advance in France in the postwar period has something to do with a reaction to a foreign threat (1992, 25–26). At least the *grands projets,* such as upgrading the telecommunications system, the nuclear program, and others are often presented as pressing responses to foreign economic, financial, or military threats. The regime's interest in science and technology has not been limited to military industries, though these were always to play an important part in the development of science in France.

Nonmilitary scientific activities supported by the monarchy included a wide range of technologies. Experimental farms were set up at Rambouillet, and a wide range of scientific and engineering projects, including shipbuilding, glass-blowing, construction trades, forestry, cabinetry, and mining, were supported by the regime (Gillispie 1980, chap. 5). In industry as well as in science, the state was heavily involved as early as the eighteenth century. This included "ownership of the tapestry looms at the Gobelins, the porcelain manufacture at Sèvres, the oriental and Turkish rug shop called the Savonnerie . . . and the establishment for upholstery and hangings in Beauvais" (Gillispie 1980, 391). Royal support was crucial in the development of a variety of new industrial applications, such as making mirrors, textiles, silks, mining, and paper. Many of the largest industrial groups of postwar France got their start with such royal support (including such giants as Saint-Gobain, which began as the Royal Manufacture of Mirrors, and Le Creusot, an early leader in mining) (Gillispie 1980, 391, 434, chap. 6). Similarly, many of the greatest schools for science and engineering date from the monarchy and stem from its interest in developing both the military and economic infrastructure to unite, enlarge, and protect the realm (Gillispie 1980, chap. 7).

The Paris Academy of Sciences was first founded to support a variety of scientific research under Colbert in 1664, and much of the great progress of European science during the eighteenth century can be attributed to researchers operating under royal support or protection, often in the state institutes. "Throughout his life, Colbert had displayed a keen interest in scientific research, even when its dividends could not be directly measured in utilitarian terms. His admiration for scientific enterprise sprang from a profound faith in the rational and precise habits of mind that were epitomized by and learned from the study of science itself" (Hahn 1971, 8–9). This attitude toward science and its role in ensuring social progress has proven profoundly enduring, despite all the changes in governmental forms that France has known. Successive regimes have often changed the forms of scientific support, but most have carried on in the consolidation and the expansion of the state infrastructure for the support of science. For example, while the *Ecole des ponts et chaussées* was created in 1775, and the *Ecole des mines* in 1783, the science and engineering corps that still dominate their professions date back even farther. State highway engineers were organized by Louis XV and his ministers in 1716. Comprehensive national planning for roads, bridges, and other engineering works began in 1747, when a central drafting office was established by the government in Paris with the purpose of assembling a set of standardized maps of the whole kingdom (Smith 1990, 659). This office later became the *Ecole des ponts et chaussées*.

The *Ecole polytechnique*, the *Ecole normale supérieure*, and the *Conservatoire national des arts et métiers* were all established in 1794; the *Institut* was created in 1795 (Hahn 1971, 288). *Polytechnique* in particular quickly became an exclusive breeding ground for France's elite science and engineering corps. Under Napoleon's centralizing administrative reforms, the *Ecole des ponts et chaussées* and the *Ecole des mines* emerged as graduate schools that recruited exclusively from among the top graduates of *Polytechnique*.

Given this long history of state-led science, there should be little wonder that scientists in France have grown accustomed to looking to the state, not to other sources, for support. In his classic work on the relations between scientists and the state in France, Robert Gilpin describes the interventionist philosophy held by French scientists during the interwar years.

> If science was to fulfill its proper function as the servant of progress and be freed from its subservience to monopolistic capitalism, there was need for a powerful and centralized organization to coordinate and give direction to French science. The goal of these scientists, therefore, was "the general organization of scientific research into a great national service of the state." Properly organized and coordinated, French science—like Soviet science—could be made to serve society. (Gilpin 1968, 158)

Indeed, the major French institutes that carry out and sponsor the country's largest scientific projects have more in common with the old Soviet model than with the U.S. model.

Gilpin begins his discussion of science policy in France with an exposition of the "technology gap" that Europeans saw developing between the United States and the national European powers in the years following World War II (1968). Nowhere were the threats of this potential relative decline seen more clearly than in France, where government officials explicitly and constantly called for increased reliance on high-technology science projects to regain and maintain France's stature in the international community. Further, given the historically great role of the state in supporting science, there should be no surprise at the large state-led effort to promote science in the first years of the Fifth Republic, as Gilpin describes. Whether the threat be economic or military, the answer of the French state has long been to reinforce and to strengthen its support for scientific research. Gilpin (1968, chap. 9) describes the special position of defense, space, and nuclear industries as guarantors of national prestige and independence; we will reinforce these findings in the argument to come.

In sum, the French state has had an extremely long history of supporting science. Relations between scientists and the state are so close that the two bodies have virtually fused in many instances, since so much of scientific research in France is conducted by people in the employ of the state operating within state institutions, or in the private sector, but at the behest of the state.

Centralization and State-Led Research Programs

Robert C. Wood argued in the early 1960s that the increasing importance of complex scientific matters in public policy would inevitably call scientists to be more involved in political matters, and he said that the "prudent political decision maker" would best be advised to bring scientists into his circle, making policy with them, expanding their influence, rather than attempting to avoid them or seeking to overrule them on political grounds (1964, 71–72). In France, scientists and engineers are constantly counted upon to assume non-scientific roles; indeed the upper reaches of the French civil service are largely staffed by graduates of the *Ecole des mines* and the *Ecole polytechnique,* as well as by the nontechnical graduates of the *Ecole nationale d'administration* (Birnbaum 1977 and 1978; Dagnaud and Mehl 1982; Rémond, Coutrot, and Boussard 1982; Ridley and Blondel 1964; Salomon 1986; Suleiman 1978). The technical and engineering training that students at these prestigious state engineering schools receive has little to do with the future administrative duties that many assume; however, the common background of administrative and technical elites leads to an extremely close working relationship.

The relationship between scientists and the state in France is more intimate than in any other advanced industrial democracy, save perhaps Japan. But this relationship is not an equal one. In France, scientists adopt a reactive posture, responding to the directions laid out by the state. State and scientific elites share a consensus about the importance of science in furthering France's international position (see Hall 1986 for a similar argument in the context of economic policy-making). In the introduction to this volume, Solingen describes such close relations as a "happy convergence." This pattern of close relations, however, comes with some very significant costs. The hyper-centralized organization of scientific research in France implies that goals must be clear. In this atmosphere, massive public efforts may be undertaken, as they have been surrounding a number of successful and impressive projects of science and engineering. However, basic science has tended not to make it to the top of the list of priorities, and has therefore suffered at the hands of state-led science in France. There are two important features of the French system of state-led science. First is the training and centralization of services inherent in it; second are the positive and negative aspects of the prioritization that this centralization requires and makes possible.

The French educational system is well known for its centralization. Along with this centralization comes a hierarchy of institutions of higher learning, ranging from the *instituts universitaires technologiques* (IUTs, or technical training centers) to the universities, and to the most prestigious of all, the *grandes écoles.* Only at the peak of this system are admissions standards rigid, and only at the peak can the French system be said to compete with the best of U.S. universities. The *grandes écoles* are one important foundation of the active science posture of the French state. Perhaps as much as any other country, and certainly more than any pluralist democracy, the state in France directly controls the training grounds of its scientists and engineers. The system of state graduate schools covers virtually the entire range of science and engineering training. Not only is it comprehensive, but it is also well entrenched in French society, politics, and economics, since it stretches back so far in time.

The educational universe has played an extraordinarily important part in the French corps phenomenon. Because of their roots as training grounds for state engineers, the best French schools have traditionally sent their graduates directly into state bureaucracies, not into private industry. Once in the higher administration, of course, many subsequently leave state service, but even this pattern of *pantouflage* is relatively recent in historical perspective. More important than whether they stay in the service of the state or work in the private sector is the corps phenomenon. Because of this distinctive and centralized system of training, French scientists and engineers constitute an identifiable body whose members share characteristics, training, and a similar view of the world. In this view, the corps is imbued with a sense of mission, and there is a wide understanding among corps members of what constitutes a problem and of how to use the state to fashion an appropriate response.

The scientific and engineering corps in France has also always held to a view that stresses the importance of rationalism and abstract principles. This has led to a faith in science and in engineering as virtually fool-proof sources of solutions to vexing problems. This does not mean that scientists and engineers always solve problems, nor that the solutions favored by them are always or necessarily the best ones available. But it does mean that this group believes that it can respond to any problem with a good solution if its members work hard enough at it, and of course if the politicians leave them alone. We will describe later the largest single example of such a state-led solution, the nuclear energy plan, but there are many others. Two examples come from telecommunications.

As any visitor to France during the 1970s or earlier can recall, the telephone system was far from a high point. But French engineers were then commanded—by the state—to experiment with electronic systems, and they installed the first digitized exchange in 1970. Today, the French telecommunications system is among the world's most advanced, and Minitel—an electronic bulletin board, electronic mail, and phone listing system—has grown in popularity and reaches a majority of French households (see Smith 1990, 658). Similarly, consider the enormously ambitious idea of replacing all the various cable networks in France with a single fiber-optics cable for telephones, television, interactive video communication, computer links, and various other professional and domestic uses (Jobert and Muller 1987, 101ff.). This belief in logical, grand, state-led solutions to social problems accounts for a fascination with technological miracles that allow the country to skip several generations of technology, moving it from the stagnation of the governmental monopoly on television service, seriously limiting potential cable-TV service, to the hypermodernism of an integrated fiber optics network not only for television, but for all sorts of electronic communications. What worked in the case of telephones could be made to work in the case of television (Cohen 1992; Jobert and Muller 1987).

The dominance of the state in all phases of scientific training and basic research gives the science policy community perhaps its most significant advantage. There is a coherence, a cohesiveness, and a direction to French science policy that is hard to find elsewhere outside of Japan and the newly industrialized countries of Asia. National priorities are defined, coordinated, and executed through a single elite, high in the state apparatus. Priorities may not always prove to be right—far from it—but at least they exist.

The centralization of the French administrative structure is unparalleled among pluralist democracies. Both in terms of its unitary structure and in the homogeneity of the social backgrounds of the elite that populate the major state agencies in France, the strength of the centralizing tendency is clear. As in other areas of French policy-making, these characteristics typify and have an important impact on science policy. Solingen argues in the introduction to this book that military involvements will lead to relative secrecy about scientific research. In the case of France, there is indeed a high level of military involvement and a similarly high level of secrecy, but as we will see with the case of the nuclear power program, this secrecy stems more from shared interests in maintaining independence from the political leadership than from national security concerns, though this is often used as a justification. As Edward Kolodziej has argued in the case of arms sales, an elite consensus in favor of such sales has meant that public debate rarely focuses on them (1987). Centralization and shared interests of the state and the scientific elite have important negative consequences for *democratic* control over science and high-technology policy in France.

The cohesive group of state bureaucrats that dominates the arms industry described by Kolodziej succeeds by convincing others that their policies serve only such national priorities as grandeur, patriotism, economic growth, technological innovation, international competitiveness, and jobs. Once this argument is accepted, questions and criticisms of the industry are muted, if not ignored. Political debate focuses on other, and most often less important, topics. The top-down creation of political consensus is key to the success of the arms industry, and it is based on strong economic arguments as well as on important symbolic images. By creating and maintaining this favorable image, Kolodziej demonstrates, the arms industry removes itself from the "heat of partisan debate and the glare of public disclosure" (1987, 402). Similar features, we will see subsequently, explain the success of the French nuclear program.

In France, a variety of state agencies support and conduct science policy. Most of these are connected to some version of a Ministry of Research. (In France, ministries are not constitutionally fixed. Any government may add, subtract, or recombine functions into ministries as it sees fit. In successive governments, the precise institutional arrangements have sometimes varied, therefore.) A list of the major French agencies with research activities is provided in table 1. A representative organizational chart of the French state in science and research is provided by table 2. Table 2 is taken from 1985; minor modifications in the titles and organizational structures in this area are common, but all governments have had a structure roughly similar to that described here.

In the twentieth century, state interest in scientific research has been accelerated and reinforced, using and expanding on the historical foundations that we have mentioned. In 1939, for example, the *Centre national de recherche scientifique* (CNRS, National Center for Scientific Research) was created (on the CNRS, see Arnaud 1979; Druesne 1975; Picard 1990; and Rouban 1987). To imagine a U.S. equivalent to this single agency, one would have to combine the National Science Foundation, the National Endowment for the Humanities, the National Institutes of Health, the Centers for Disease Control, and those faculty members in universities that regularly conduct research under contract with these federal agencies. In 1988, the CNRS employed over 11,000 full time researchers, as noted in table 1.

The importance and distinctiveness of the CNRS can be noted by consid-

TABLE 1. Principal Research Organizations in France

Unit	Number of Researchers	1988 Expenditures (millions FF)
CNRS (National Center for Scientific Research)	11,254	10,331
INSERM (National Institute of Health and Medical Research)	1,939	1,825
INRA (National Institute of Agronomic Research)	1,668	2,454
CEA (Atomic Energy Commission)	1,644	2,989
CNES (National Center for Space Studies)	1,063	739
Pasteur Institute	850	345
Other important research centers		
AFME—French Agency for Energy Management		
ANVAR—National Agency for the Application of Research		
CEMAGRAF—National Center for Agricultural Machinery, Rural Engineering, Water and Forestry		
CNET—National Center for Telecommunications Studies		
CRITT—Center for Research, Innovation and Technology Transfer		
INED—National Institute for Demographic Studies		
INRETS—National Institute of Transport and Transport Safety Research		
INRIA—National Institute of Computerization and Automation Research		

Sources: Economic and Social Council; CREST, 1990:32–33.

TABLE 2. 1985 Organizational Chart of French Government in Science and Research

- Ministry of Research and Technology
  - Delegation for Information, Communication, and Scientific and Technical Culture
  - The Scientific and Technical Office (*mission*)
    - Division for Basic Research
    - Division for Science and Techniques of Industrial Systems
    - Division for Research in Social, Economic, and Cultural Domains
  - Directorate-General for Research and Technology
  - Agency for the Organization and Promotion of Research
    - Office for Scientific Employment and Research Structures
    - Office for the Promotion of Research and Innovation
  - Agency for Research Finance
    - Office of the Budget
    - Office of Finance
  - The Program Office
    - Department of Planning and Programs
    - Bureau of Program Synthesis
    - Department of Mobilization Programs
    - Department of Technological Development Programs
    - Department of Laboratories
    - Department of Basic Research
    - Department of Scientific Infrastructure
  - Mobilizing Programs
    - Committee on the Development of Biotechnologies
    - Committee for the Development of the Electronics Sector
    - Committee for Scientific Research and Technological Innovation for the Third World
    - Committee for Research on Employment and the Improvement of Work Conditions
- Joint Bureaus with the Ministry of Industrial Redeployment and Exterior Commerce
  - Center for Planning and Evaluation
  - Office of Industrial Strategies and Statistics
  - Directorate-General for Regional Development and Industrial and Technological Environment
  - Delegation for International Affairs
    - Office of Bilateral Affairs
    - Office of Multilateral Affairs
  - ANVAR—National Agency for Research Evaluation (*valorisation*)
  - CNRS—National Center for Scientific Research
  - Interdisciplinary Programs
    - Energy science and raw materials research
    - Forecasting and observation of volcanic eruptions
    - Research on the scientific bases of medicines
    - Environmental research
    - Oceanographic research
    - Materials research
    - Research on technology, work, employment, and life styles
  - National Institute for Universal Sciences
  - National Institute for Nuclear and Particle Physics

Source: Ministry for Research and Technology internal memoranda.

ering one thing that the CNRS does not do. It does not teach a single student. It is as if science in France takes place in a single, vast Rand Corporation that is entirely government owned and operated. The universities are generally separate from the CNRS, though there are many joint laboratories and many CNRS centers are physically housed on or near university campuses. Indeed, the majority of CNRS laboratories are joint or mixed labs, usually associated with universities, and many CNRS researchers take on advanced students from the universities. Rarely, however, would they encounter a student at the undergraduate level. In the French system, the universities would be unable to offer serious research training without linkages with the CNRS. Students working on doctorates essentially must leave the regular university in order to acquire research skills available only in the CNRS laboratories, and the CNRS personnel in these laboratories maintain their distinct privileges from the rest of the university faculty. France has never subscribed to the U.S. taste for intentionally mixing teaching and research in the comprehensive research university, although some French scientists argue that France should move in this direction. Even in the prestigious *grandes écoles,* emphasis is mostly on teaching. The separation between teaching and research in the French higher education system is one of its most distinctive characteristics, and one of the country's biggest problems for training scientists and engineers.

The CNRS has also tried to "modernize" its somewhat archaic feudal structures (Picard 1990, 263). One weakness of the CNRS has traditionally been its segmentation between traditional scientific disciplines. The comprehensive research university on the U.S. model could not differ more from the CNRS, which is organized into dozens of small research laboratories, each physically separated from the other, and each with its own budget, research agenda, and often controlled by one or a few senior scholars (sometimes known as *patrons* or *mandarins*). Such an organization has made especially difficult the kind of interdisciplinary collaboration and cross-fertilization that modern science increasingly requires. There are of course efforts to address these problems, but they remain important features of the French research establishment.

This complex teaching-research universe in France resembles, in fact, a caste system. In higher education, the IUTs constitute a woefully underfunded system of technical and/or vocational training schools. They may grant the Diplôme Universitaire de Technologie degree, but may not grant bachelors or graduate degrees. The regular universities are equally poorly funded, but they do purport to offer the panoply of standard science and humanities degrees at both the undergraduate and graduate levels. Many professors in the university system have research agendas, but are so overworked by teaching duties and underfunded, if funded at all, for research that relatively little significant research is accomplished in the French university. The *grandes écoles* are of course the elite track. They enjoy relatively high funding and the professors in them are better paid, teach fewer courses, and generally turn out the graduates who go on to become the famous scientists and engineers of France. Still, compared to a U.S. Georgia Tech or MIT, little research takes place inside the French elite *grande école*. In France, the complex caste system is complete only with the addition of the CNRS and the other elite state institutes devoted exclusively to research, such as the National Institute of Health and Medical Research (INSERM) or the Pasteur Institute. So not only is there more distinct hierarchy of universities in France than elsewhere, but there is also a greater separation between units that conduct research from those, even very prestigious, schools that mostly focus on teaching.

There are a number of weaknesses associated with this state-led system of science. Gilpin (1968, chap. 4), for example, argues that administrative rigidities produce lack of communication across ministries, except at the highest levels. Papon discusses how the various research arms of the French government, which are indeed impressive, lack any coherence since they are all controlled by the various specialized ministries (1978, 219–20). Like the train traveler hoping to go from Cherbourg to Brest, but who finds that the LeGrand Star has made it necessary for him to travel via Paris, the researcher hoping to coordinate efforts with his counterpart in another ministry often finds tremendous institutional hurdles that make it necessary to have coordination at the highest levels of the administrative structure. In other words, as all train routes go through Paris, all administrative decisions are made through coordination at the top. However, the connections are often not as easy. So state-led science can be too big for effective coordination.

A significant weakness of the French state-led system of science continues to be the separation of universities from research institutes and *grandes écoles*. This means that those who create the knowledge are far removed from those who pass it along. Further, engineering training in France has traditionally focused on learning skills and the content of scientific research, but not the research process itself. Finally, the multiplicity of *grandes écoles,* each specialized in its own area of research, has made it difficult for French scientific establishments to adapt to technological innovations. For example, there was no way to major in a French university in chemical engineering throughout the nineteenth century, even while chemical innovations were at the center of all technological progress at that time. As one educational administrator said in a 1984 interview for another project (Baumgartner 1989), "if your school is called 'the National School of Chemistry,' you cannot change its focus to computers." While the multifaceted U.S. universities can create new departments or new institutes over time, the rigidities inherent in the specialized French system have hindered their ability to react to innovations. Of course, examples of bold innovation exist, and one only need to look at the *Ecole des mines,* whose graduates dominate many areas of French administration and research, especially those areas associated with energy, to see that a school can change its focus as its original charter becomes obsolete. Still, the small size and great specialization that characterizes high level training institutes in France constitutes an important problem for its scientific infrastructure. The government is aware of the problems of this hierarchical system, and a recent proposal adopted by the Council of Ministers calls not only for a doubling in the numbers of engineers trained in France in only a four year period, but also that large numbers of these future engineers should be trained in the universities rather than in the *grandes écoles,* thus providing the first break with the traditional system of complete separation (Courtois 1990).

State-led science can be very successful. The Messmer Plan to develop nuclear energy after the oil shocks of 1973–74, which we will turn to subsequently, is a stunning testimony to the success and effectiveness of the state-led approach. But there are other examples as well: consider solar energy research.

The oil shock of 1973–74 excited more than just the partisans of nuclear energy in France. Faced with a national crisis, researchers in all areas of energy production pressed their various solutions. At the National Center for Scientific Research (CNRS), a lone researcher named Félix Trombe had been engaged in solar energy research as early as the 1950s and had established a field research site in the Pyrenees, which had been roundly criticized for its cost and for its lack of readily evident, usable results. With the onset of the crisis in 1974 and the call in France for a general "scientific mobilization" in the search for alternative energy forms, the CNRS seized upon solar energy with great enthusiasm. The reasons for this were partly idiosyncratic, since it already owned the Pyrenees center and could use this as a justification and as an answer to its past critics, partly philosophical, since solar energy provided a unique opportunity to further the CNRS agenda for increased interdisciplinary cooperation in research, and, finally, partly environmental, since solar energy promised none of the potential hazards of nuclear power. A new CNRS interdisciplinary research group was formed and its first budget was a respectable 23 million francs. The Pyrenees center was brought fully on line, and an additional laboratory was established in the southern Alps for the development of amorphous silicum panels.

Excitement ran high enough for the Under-Secretary for Research, Jacques Sourdille, to encourage the CNRS group to cultivate contacts with public and private enterprises that might benefit from the research. In a state-led system of science, of course, a suggestion by a government official placed in the overseeing ministry is tantamount to an order. What Sourdille had in mind was more than just polite interchange of information but rather an integration of industry into the research. Eléctricité de France (EDF) was commissioned for the manufacture of the solar units, Fives-Lille et Cail for the heat conversion system, and Saint-Gobain for the mirrors (Saint-Gobain was, of course, the premier French manufacturer of mirrors, this dating from the Royal Manufacture of the eighteenth century: see previously). The result, as Picard notes (1990, 265), was "a beautiful technical demonstration of the CNRS's savoir-faire in automated heliostats [solar panels that follow the sun's movement]." A technological triumph, the CNRS solar energy project was, however, a spectacular commercial failure. Even worse than the Concorde, which at least has operated commercially for years even though thoughts of recovering its large development costs have long been abandoned, this project never recovered any of the state's investment (for greater detail on the CNRS solar energy example, see Picard 1990, 264–69).

As discussed previously, the demonstration of technical prowess does not always lead to commercial success, but in state enterprises there is often more emphasis on the former than on the latter. In France as a whole, state agencies are responsible for a greater proportion of R&D activity than private firms, in contrast to other western countries.

Table 3 shows the relatively impressive record of the French state in supporting R&D expenditures. It shows that as a proportion of GDP, the French state is the most active of all the European Community members in supporting R&D expenditures. Further, this activity has increased during the 1980s. However, France is not a world leader in the total percentage of GDP spent on R&D when one includes private sector expenditures, because the French private sector has traditionally spent little in this area. In 1985, France had only 4.3 research scientists per 1,000 members of the work force, compared with 5.2 in West Germany, and 6.5 in the United States. Figures for engineers show an even greater relative lag for France compared to its major competitors (Commission du bilan 1981, 223; European Community 1990, 19).

TABLE 3. Total Public Expenditure on R&D as Percentage of GDP

	France	FRG	UK	Italy
1980	1.12	1.15	1.08	.46
1981	1.31	1.15	1.31	.65
1982	1.32	1.21	1.34	.64
1983	1.41	1.14	1.33	.70
1984	1.46	1.11	1.35	.76
1985	1.47	1.15	1.32	.78
1986	1.44	1.10	1.24	.85

Source: European Community 1988, 50–51.

Part of the reason for the large state role in supporting the national R&D effort is, of course, the large percentage of French R&D devoted to defense. In addition, French businesses have become more active in supporting research and development, and this trend can be expected to continue with the consolidations and enlargements taking place in every field of European business. For example, in 1961, public R&D expenditures as a percentage of GDP in France were over 1.5 percent, while the figure for the entire private sector was only just above 0.6 percent. By 1977, private expenditures on R&D had increased only to approximately 0.7 percent of GDP, but since then they have increased steadily. In 1989, public expenditures were approximately 1.3 percent of GDP, while private effort was about 1.0 percent. In other words, during the period since 1967, the ratio of public to private spending on research and development declined from approximately 2.5 : 1 to about 1.3 : 1 (European Community 1990, 4). Private business in France is devoting more and more of its own resources to research and development, but as a whole the country remains strongly affected by its statist past.

The relative lack of scientific initiative within the French firm is part and parcel of a national perspective that looks first to the state for initiatives. Of course, a number of French firms are leaders in aggressive R&D activities, but as a whole, Gillispie's description of eighteenth-century practices remains true today: "French entrepreneurs habitually looked first to the state rather than to the financial markets or to private savings to provide the capital they clearly needed" (1980, 388). This tradition has carried through the centuries in France. Scientists or industrial concerns look first to the state today when they seek support for research and development expenditures.

Nuclear Physicists and the State

Nuclear energy policy in France is a prime example of the shared interests of those elites within and outside the state in the depoliticization of the debate, and of their success in convincing large majorities of the broader political elite that "the experts" know best and should be in control. While France had developed its indigenous nuclear power industry in the 1950s and 1960s, it was not until the 1973 oil crisis that the industry took off. In one of the largest public policy decisions since the war, French leaders decided in 1974 on the Messmer Plan. This was the proposal to replace as much of the country's electrical generating capacity as possible with nuclear power plants. After important (and, among elites, controversial) decisions to abandon the indigenous French design and to standardize construction of a series of identical plants using a Westinghouse pressurized water (PWR) design, the French electrical company EDF went ahead with a massive building plan. Twelve years later, fifty-three plants were in operation, and the proportion of the nation's electricity generated by nuclear power had risen from less than eight in 1974 to over eighty percent in 1990. Figure 1 shows the scope of France's Messmer Plan, and the success of its implementation.

The Messmer Plan was the result of elite consensus, not public opinion. Opinion polls show that the French public is not particularly hostile to nuclear energy, but neither has it always favored it. Further, public opinion in France is not very different from public opinion in neighboring countries with much less ambitious nuclear programs and where nuclear power has been much more controversial. Ronald Inglehart (1984) shows, for example, that French opinion in 1983 was actually less supportive of nuclear power than U.S. opinion, and that it fell in line with most of its European neighbors. National policy in France has been pronuclear because of an elite consensus and because of proponents' success in keeping the question away from the political agenda, not because of public opinion differences.

Of course, there have always been opponents to nuclear power in France. But the most significant defining characteristic of the opposition has been its complete lack of success. EDF was particularly concerned with public opinion during the 1970s when it was expanding its program so rapidly, and tracked public opinion regularly. Its researchers found that antinuclear sentiment was generally related to the left-right cleavage in French politics, as in most other countries (Ansel, Barny, and Pagès 1987). Communist party supporters were particularly likely to be antinuclear, but of course the Party leadership was firmly in support of the program. Similarly with the Socialist party, at least after its initial hesitations. While the voters were split at best over the technology, leaders of the two leftist parties were uniform in their public support of nuclear power. In sum, the nuclear industry has been depoliticized in France, not because the mass public is particularly in favor of it, but because of an elite-level consensus not to exploit the issue.

Dependent on the state for research funds, sharing with the state elite (of which they are virtually a part) a view of the country's problems that stresses large-scale state intervention to solve them with technological solutions, and reacting with enthusiasm to the state's decisions to rely on science to achieve its goals, French nuclear scientists and engineers illustrate many of the merits, as well as many of the costs, of the French national style of state-science relations.

As in other areas, France's preeminence in the scientific discoveries associated with atomic energy during the nineteenth and early twentieth centuries have more recently taken a back seat to the enormous civil engineering project of *le tout nucléaire,* the goal of producing virtually all of France's electricity in nuclear power plants. Lawrence Scheinman has provided an excellent analysis of the genesis of French nuclear power, mostly in its military applications, during the Fourth Republic (1965). Similarly, Bertrand

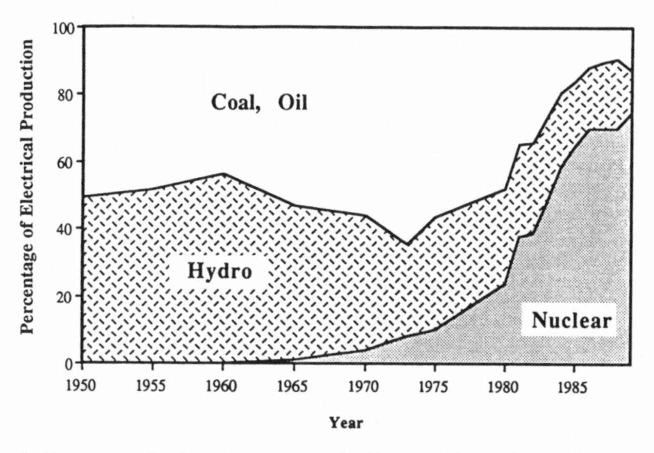

Fig. 1. The growth of nuclear power in France. (From Commissariat de l'Energie Atomique 1986.)

Goldschmidt (1987) has provided the perspective of one of those involved during World War II in the effort by the French to develop their independent *force de frappe,* considered at the time and since then to be essential to the maintenance of the country's independence and continued national prestige. What role have French nuclear scientists and physicists played in the struggles for arms control, human rights, or international equity, suggested by Solingen at the beginning of this volume as important potential roles? In these fields, French physicists have played no particular role. They *have,* however, played a major role in the technical and development work necessary to maintain France's position in the world.

Scheinman (1965) described how even at the inception of the French nuclear program, it avoided partisan battles. Many of the premier nuclear scientists at the time were members of the Communist party, which supported the program. Within the national electrical company, the Communist-led Confédération Générale du Travail (CGT) has long been the dominant union, and it has firmly supported expansion of the nuclear program (Picard, Beltran, and Bungener 1985). One important battle that in other countries has put unions and the political parties associated with them at the center of the nuclear power debate has to do with the future of coal. In the United States, Great Britain, and especially West Germany, coal miners were an important force. In France, the decline of the northern coal mines took place while the major union that represented coal miners, the CGT, was the same as that which was strongest in the nuclear area and within EDF as a whole. The Communist party, associated with the CGT, had nothing to gain by focusing on a battle pitting two of its own bases against each other. Similarly in the case of oil, another battle was avoided in the French case. With no domestic sources of oil, few politicians were attached to the industry in France, and the leaders of the nationalized oil companies never opposed the expansion of nuclear power in France as oil interests have done in other countries. So with neither coal nor oil interests mobilized against them, the expansion of nuclear power did not arouse partisan mobilizations in France. Nuclear power has been seen in France as the guarantor of national independence in the military realm, and almost in the same terms for civilian nuclear power.

France's success in displacing oil, in implementing a massive public policy program toward nuclear power, in becoming a major exporter of electricity, and in maintaining the costs associated with nuclear power better than any of its neighbors, have all reinforced the elite level consensus that made the program possible. Nuclear power in France is the subject of virtually complete acceptance. Scientists and state elites, acting in concert, have depoliticized a major national policy. In this area, the strength of the French state working with a consensual scientific community stands in stark contrast to that of most other Western democracies.

The international dimension in the French nuclear program has been important from its inception. However, its greatest impact has been in providing the proponents of the program with powerful arguments that they could use in the domestic context in order to develop a consensus. The need to maintain and enhance France's position as a world power was obviously a strong argument for France's decision to develop the independent striking force. External economic "threats" were also key to the justification of France's decision to develop the civilian nuclear program. International prestige and competitiveness are not partisan issues in France. Scientists capable of convincing state authorities that their work tends only to further these goals can by-pass political and partisan bickering completely. Just as Kolodziej (1987) described in the case of the *Délégation générale de l'armement* and the arms industry, proponents of the nuclear program have been successful in linking their policy with these consensual goals. Further, executive branch officials charged with implementing large programs want nothing more than to avoid politically charged debates in Parliament. Therefore, they are pleased to surround the nuclear power program with the positive image from which it benefits today. So international relations have historically played an important role in both the military and civilian aspects of France's nuclear program, but they play an indirect role. They offer powerful arguments that are very useful in domestic political quarrels.

In contrast to that in many other countries, the French nuclear program works, and works well. There are few political problems associated with it, and technical problems are fewer than elsewhere. French leaders of the program take justifiable pride in these facts, and see them as important elements in developing France's international reputation for high-technology industries. International prestige is a clear goal of the nuclear program.

Another important goal of the nuclear program was not very important until 1973, but it transformed the French program and gave it its present grand scope. The main goal of the French nuclear industry shifted from focusing on the development of national technologies, and, in fact, the French design used until then was scrapped. The U.S. Westinghouse PWR design was chosen, and emphasis suddenly shifted toward engineering skill, organization of a successful industrial complex, and the most rapid displacement possible of imported oil for energy. So nuclear power became, and remains today, the key to France's energy independence. The goal of demonstrating the superiority of the domestic design of Gas Graphite Reactors was abandoned with remarkable readiness when a more important international threat was perceived. Scientific research, represented in the development of the French design, lost out quickly when put into competition with an engineering goal.

Still another major international goal associated with the nuclear program is also new. In the 1980s, France became a major exporter of electricity. France began the decade with net imports of just over 3 billion kWh of electricity. By 1981, however, it was exporting almost 5 billion kWh; 24.7 billion in 1984; 29.7 billion in 1987, and 45.7 billion in 1990 (Electricité de France 1988; French Embassy 1991). With virtually every country reeling from overpredictions in the rate of increase in demand for electricity, the fact that France now exports over 15 percent of the electricity it produces is a major advantage both in terms of foreign trade balances and in providing a justification for the huge construction projects of the 1970s and 1980s. In the deregulation of the European energy markets to follow 1992, France expects to use its competitive advantage, which stems from its highly efficient nuclear plants, to improve its foreign trade.

The strength of these images and the link between civilian and military nuclear power makes France peculiar among democracies in that the military aspect of the fast breeder program, the production of plutonium for military uses in an otherwise civilian plant, is not controversial. In the United States and in other countries, great efforts have been taken to maintain the separateness of the civil and the military programs. This is rarely a concern in France. Both are seen to enhance national independence, prestige, and power. France's two fast-breeder reactors, the Phenix and the Super-Phenix have been the subject of considerable controversy, but the questions have been more economic than military. Indeed, the most significant questions have been raised about the financial need for the fast-breeder technology, even as France became the only country in the world to build a full-scale fast-breeder and to connect it to its power network. The Super-Phenix, far from being the symbol of French technological prowess and the first of a new generation of reactors, as it was planned, has instead turned into a technical and financial nightmare, operating only for less than two years and closed for a variety of technical reasons for most of the time since it was first completed. While what the French call the "classical nuclear" industry has been a relative success, the fast-breeder technology has shown the pitfalls of the French emphasis on high-prestige, technologically complex projects with risky commercial futures.

French nuclear officials are also proud of their ability to handle wastes, their retreatment plant at La Hague, their underground vitrification facilities, and the lack of controversy surrounding waste policies even allow France to be one of very few western countries that actually imports nuclear waste, in particular from Japan. From the export of electricity to the importation of spent fuel and waste, France's nuclear power industry has served all three goals of national prestige, energy independence, and, increasingly, commercial success and foreign trade benefits. French officials therefore see their nuclear program as a great policy success in the sense that all three of these goals have been met, not only the first one as in some other cases.

French Science Policy in the 1990s

The stated goals of French science policy as the country entered the 1990s were sixfold, as outlined in the 1989 finance bill (CREST 1990, 1). These were to give priority to industrial research, to maintain and increase the resources allocated to basic research, to promote the linkage between basic and applied research, to encourage industry to increase its research activities, to support "major" technological programs, and to foster European cooperation in research efforts. Thus one can see that the French state has become sensitive to some of the shortcomings that we have identified with its state-led style of science policy, in particular the relative lack of basic research in a system dominated by applied research and engineering and the relative lack of research activity inside industry itself.

The French carry with them into the 1990s a keen awareness of these weaknesses and of their position in the world relative to their main competitors. Government policy explicitly lays out a catch-up strategy designed to strengthen French performance in the areas of weakness that we have noted. But the approach is still a remarkably state-led one. In a policy statement made on 14 December 1988 by Hubert Curien, the minister for research and technology, the French state-led strategy was made unmistakably clear.

> The objective is to match what our neighbors are currently spending and to raise national R&D expenditure to three percent of GDP by consolidating those areas in which we excel, correcting our weaknesses and creating the conditions for better dissemination. This effort must be accompanied by a determination to restore balance to our structures for the financing and execution of national research expenditure. There is one precondition which can and must be fulfilled: France's desire for strategic autonomy has led it to develop high-quality military research,

> but it is also necessary to improve the synergy between military and civil R&D programs and to do more, along the lines of what is happening in the United Kingdom, to apply the results of military programs to industry. . . . Companies must also significantly increase their research effort. A number of measures proposed are designed to encourage them to go along that road. Consequently, the increase in national research expenditure cannot be divorced from the increase in the share to be funded by the State. Priority will be given to the civil research budget over the coming years. The size of the budget will be increased so as to ensure that France, wherever possible, is placed on a par with its main competitors in 1993. (CREST 1990, 5)

The French state's scientific program for the 1990s stresses a number of specific projects in addition to the general principles outlined by Curien. These include well-developed programs in information technology, molecular genetics, AIDS, medium-sized equipment acquisition, the construction of large-scale scientific research facilities, and improvements in technology transfer and dissemination. The French state's support for general research in civil aeronautics, space, nuclear power, telecommunications, and other areas of traditional strength will continue at consistently high levels. The techniques that the French will use to manage and channel research in all these areas will continue to look very familiar. They will use a host of subsidies, grants, incentives, tax credits, and other means of direct and indirect intervention to shape the country's scientific research agenda, always with specific overall goals in mind. The Mitterrand administrations have consistently provided high funding levels to education and to research (Salomon 1987), and these priorities continue in the 1990s.

There are also European Community research programs of scientific import, and France actively cooperates with these efforts and often takes the lead in shaping them. The Community operates its own research through a network of four Joint Research Centers (JRCs). These centers specialize in work on nuclear fission, environmental protection, technical standards, and nuclear safety. However, together these centers only employ 2,200 people, of which only 700 are engaged in active research. A much larger proportion of the community's spending on scientific research goes toward "shared costs." In this, the community contributes 50 percent of the cost of research that it approves that is carried out in institutes, universities, and both private and public companies across all member-states. Many of the multinational community programs are carried out on a shared-cost basis. Finally, some of the community's research activities take the form of "coordinated action," in which the community defrays the administrative costs of coordinating a comprehensive research project, but does not fund the research itself.

While the community claims that in some sectors of scientific and technology research, such as nuclear fusion or particle physics, Europe remains at the forefront of world leaders, a relative decline is clear, and increasingly so in important strategic sectors such as electronics, information technology, biotechnology, or materials technology. The community has adopted the view that one of the principal causes of this relative decline compared to the United States and Japan is not one of lack of funding or ability but rather extreme fragmentation. It believes that resources are too dispersed, research teams are too isolated, research is poorly coordinated, information is poorly diffused, and national programs duplicate each other (European Community 1987).

The community has developed a framework program for scientific research and development. The second framework program, valid from 1987–91, calls for specific resources to be devoted to eight areas. The program is not meant to centralize the carrying out of research at the community level, but to act as a strategic coordinator for otherwise disparate and dispersed research programs. Some funding goes as well into an incentive structure meant to increase research activity in certain scientific domains. The 1987–91 framework program is displayed in table 4.

The Community intends the framework program to accelerate the establishment of a truly European scientific and technical area, "an authentic European Community of research and technology, an indispensable ingredient of the 'large market without frontiers' which the community aims to establish between now and 1992" (European Community 1987, 5). While France supports this line, both in terms of moral and material support, it is safe to say that France does not regard the European Community as a substitute for the scientific undertakings within its own house.

France also does not view the European Community as the sole or exclusive framework for European and other international scientific cooperation and coordination. France took the lead in establishing EUREKA, designed to place Europe at the forefront of world-wide technology in a variety of areas ranging from lasers to robotics. French support for EUREKA amounted to 740 million francs in 1988. France has also played a major role in the development of European scientific agencies and specific large-scale facilities, such as the European Organization for Nuclear Research (CERN), the Large Electron Positron Collider (LEP), the European Space Agency (ESA), the European Molecular Biology Organization (EMBO), the European Synchrotron Radiation Facility (ESRF), the Institute for Extremely High Frequency Radio Astronomy (IRAM), and many others. Moreover, France sustains many bilateral R&D cooperation agreements, especially with the United States, Japan, and Germany (cf. CREST 1990, 26–28). Thus, European ventures will clearly be an important part of the French research and science policy for the decades to come, but the country is not yet at a stage where one would argue that the European programs are more important than the national ones; far from it.

The Happy Convergence of French Scientific and Administrative Elites

The French political system is a highly centralized one. In France, indeed, all roads (and railroads) *do* lead to Paris. We have seen that science policy is no different. A state-led strategy for science or anything else requires centralization in order to effectively coordinate any broad plan. The postwar period in France, in particular, was characterized by a vigorous return to planning and state intervention in order to reconstruct the economy. The planning played a role in the *Trentes glorieuses,* or the roughly thirty-year period from 1944 to

TABLE 4. European Community Framework Program, 1987–91 (amounts in millions ECU)

Amount	Program	Amount	Subprogram
479	Quality of life		
		80	Health
		65	Radiation protection
		334	Environment
2,465	Information and communications		
		1,790	Information technologies
		550	Telecommunications
		125	Transport and new services
989	Industrial modernization		
		460	Manufacturing technologies
		240	Advanced materials
		72	Raw materials and recycling
		217	Technical standards, measurements, reference materials
310	Biological resources		
		140	Biotechnology
		105	Agroindustrial technologies
		65	Agricultural competitiveness
1,752	Energy		
		542	Fission, nuclear safety
		1,000	Controlled fusion
		210	Nonnuclear energy
80	Third World development		
80	Marine resources		
		50	Marine science and technology
		30	Fisheries
325	European scientific cooperation		
		205	Stimulation
		30	Use of major installations
		25	Forecasting, assessment
		65	Dissemination

Source: European Community 1987.

1975, when the French economy, largely at the behest of the state, was transformed from a backward, small-village, small-plot society, whose industry was sluggish and whose innovations never seemed to take hold, to a modern, contemporary, glassy, glossy country. The confrontation of old France with new France during this period is depicted in Jacques Tati's *Mon oncle,* wherein France is suddenly overrun by cars and highways and new gadgets from telephones to microwaves. The *Trentes glorieuses* hearken back to the state activism of the *ancien régime* or to the nineteenth-century period of planning under Legrand (Gilpin 1968; Price 1965). In between, of course, lie numerous periods of stagnation, conservatism and, worse, Malthusianism. Throughout, the state has been at the center of French science and research, be it during periods of rapid innovation or of stagnation.

In this universe, we have seen that the relationship between scientists and the state in France is, and has long been, harmonious, a happy convergence. The French state has for many centuries been active in the promotion of scientific innovation, both through a view of science's role in furthering the glory of France and through a long history of domestic and foreign intervention by the French state in which science was, and remains, useful. The relationship between scientists and the state in France is therefore more intimate perhaps than in any other democratic, pluralist country, save Japan, as we have argued. But this relationship is not an egalitarian one. In France, scientists react to the directives of the state. The direction that science follows is orchestrated by the French state, which assumes a very active, interventionist posture. This generalization holds true for a host of scientific policy domains, ranging from the nuclear *force de frappe* and the vast high-technology French armaments industry to the extensive, comprehensive French nuclear energy development program.

France's distinctive international position, the centralization of its administrative structures, and the shared backgrounds of its scientific and administrative elites ensure that science policy is and will remain a central concern of the state. These factors coincide to produce a scientific community much more dependent on the state than in many other countries, but also much more easily mobilized by the state to solve important domestic and international problems. The greatest triumphs of French science have come when large-scale, state-led projects have succeeded, as in the case of the nuclear program, the space program, or high-speed trains. This same process, however, that has led to an emphasis on large infrastructure projects has also caused a number of problems. The Plan Calcul, the state's plan to renovate the country's computer industry, restricted the importing of computers in France for years without leading ultimately to a competitive French industry. Similarly, the lack of emphasis on basic research and the dependence of the private sector on public sector research has meant that French science cannot be at the forefront in all areas at all times. In those areas that the state makes priorities, there are some stunning success stories and some notable failures. However, in those areas not identified by the state as priorities, there are very few successes. In sum, the state rather than private industry plays a key role in defining the priorities of science in France. These priorities tend to be linked closely with relatively immediate goals of economic and military competitiveness, as well as with the over-arching goal of national prestige, or *grandeur.*

REFERENCES

Ansel, Philippe, Marie-Hélène Barny, and Jean-Pierre Pagès. 1987. "Débat Nucléaire et théorie de l'opinion." *Revue Générale Nucléaire* 5 (September-October): 451–59.

Arnaud, Robert. 1979. *L'arbre à deux branches: La grande aventure du CNRS*. Paris: Presses de la Cité.

Baumgartner, Frank R. 1989. *Conflict and Rhetoric in French Policymaking*. Pittsburgh, PA: University of Pittsburgh Press.

Birnbaum, Pierre. 1977. *Les sommets de l'Etat*. Paris: Seuil.

Birnbaum, Pierre. 1979. *La classe dirigeante française*. Paris: Presses universitaires de France.

Boulloche, André, Klaus Richter, and Kenneth Warren. 1976. *The Sciences and Democratic Government*. London: Macmillan.

Brooks, Harvey, and Chester Cooper, eds. 1987. *Science for Public Policy.* Oxford: Pergamon.

Brun, Gérard. 1985. *Technocrates et technocratie en France, 1914–1945*. Paris: Albatros.

Calon, Michel, ed. 1989. *La science et ses réseaux*. Paris: Editions de la Découverte.

Caullery, Maurice. 1948. *La science française depuis le XVIIe siècle*. Paris: Armand Colin.

Cohen, Elie. 1992. *Le Colbertisme "high tech": Economie des Télécom et du Grand Projet*. Paris: Hachette.

Commissariat à l'énergie atomique. 1986. *Rapport Annuel 1986*. Paris: Commissariat à l'Energie Atomique.

Commissariat Général du Plan. 1989. *La science, la technologie, l'innovation: Une politique globale*. Paris: La documentation française.

Commission du bilan. 1981. *La France en mai 1981: l'enseignement et le développement scientifique*. Paris: La documentation française.

Courtois, Gérard. 1990. "Le gouvernement veut doubler les flux d'ingénieurs en quatre ans." *Le Monde,* 28 September.

CREST. 1990. *Comparison of Scientific and Technological Policies of Community Member States: France*. Luxembourg: Office for Official Publications of the European Communities.

Dagnaud, Monique, and Dominique Mehl. 1982. *L'Elite rose*. Paris: Editions Ramsay.

Deheuvels, Paul. 1990. *La recherche scientifique*. Paris: Presses universitaires de France.

Dhombres, Nicole. 1989. *Naissance d'un pouvoir: Sciences et savants en France, 1793–1824*. Paris: Payot.

Druesne, Gérard. 1975. *Le centre national de la recherche scientifique*. Paris: Masson.

Electricité de France. 1988. *Résultats techniques d'exploitation, 1987*. Paris: Electricité de France.

European Community. n.d. *L'Europe et les nouvelles technologies*. Luxembourg: Office of Publications of the European Communities.

European Community. 1987. *Research and Technological Development for Europe*. Luxembourg: Office of Publications of the European Communities.

European Community. 1988. *Basic Statistics of the Community*. Luxembourg: Office of Publications of the European Communities.

European Community. 1990. *Comparison of Scientific and Technological Policies of Community Member States: France*. Luxembourg: Office of Publications of the European Communities.

Feigenbaum, Harvey B. 1985. *The Politics of Public Enterprise*. Princeton: Princeton University Press.

Fox, Robert, and George Weisz, eds. 1980. *The Organization of Science and Technology in France, 1808–1914*. Paris: Editions de la maison des sciences de l'homme.

French Embassy. 1991. "Nuclear energy still primary source of electricity." *News From France*, 21 February.

Gillispie, Charles Coulston. 1980. *Science and Polity in France at the End of the Old Regime*. Princeton: Princeton University Press.

Gilpin, Robert. 1968. *France in the Age of the Scientific State*. Princeton: Princeton University Press.

Gilpin, Robert, and Christopher Wright, eds. 1964. *Scientists and National Policymaking*. New York: Columbia University Press.

Goldschmidt, Bertrand. 1987. *Les pionniers de l'atome*. Paris: Stock.

Hahn, Roger. 1971. *The Anatomy of a Scientific Institution: The Paris Academy of Sciences, 1666–1803*. Berkeley: University of California Press.

Hall, Peter. 1986. *Governing the Economy*. New York: Oxford University Press.

Holmes, Peter, and Margaret Sharp, eds. 1989. *Strategies for New Technology: Case Studies from Britain and France*. New York: Allan.

Inglehart, Ronald. 1984. "The Fear of Living Dangerously: Public Attitudes Toward Nuclear Power." *Public Opinion* 7: 41–44.

Janicaud, Dominique, ed. 1987. *Les pouvoirs de la science: un siècle de prise de conscience*. Paris: Vrin.

Jobert, Bruno, and Pierre Muller. 1987. *L'Etat en action*. Paris: Presses universitaires de France.

Kolodziej, Edward A. 1987. *Making and Marketing Arms*. Princeton: Princeton University Press.

Latour, Bruno. 1989. *La science en action*. Paris: Editions de la Découverte.

Long, Theodore, and Christopher Wright, eds. 1975. *Science Policies of Industrial Nations*. New York: Praeger.

Ministère de la Recherche. 1988. *La France dans L'Europe de la science et de la technologie*. Paris: La documentation Française.

Olivennes, Denis, and Nicolas Baverez. 1989. *L'Impuissance public: L'Etat, c'est nous . . .* Paris: Calmann-Lévy.

Papon, Pierre. 1978. *Le pouvoir et la science en France*. Paris: Editions du Centurion.

Paul, Harry W. 1985. *From Knowledge to Power: The Rise of the Scientific Empire in France, 1860–1989*. Cambridge: Cambridge University Press.

Pestre, Dominique. 1984. *Physique et physiciens en France, 1918–1940*. Paris: Editions des archives contemporaines.

Picard, Jean-François. 1990. *La république des savants: La recherche française et le CNRS*. Paris: Flammarion.

Picard, Jean-François, Alain Beltran, and Martine Bungener. 1985. *Histoires de l'EDF*. Paris: Dunod.

Piganiol, Pierre. 1987. *La recherche mal menée?* Paris: Larousse.

Piganiol, Pierre, and L. Villecourt. 1983. *Pour une politique scientifique*. Paris: Flammarion.

Price, Don K. 1965. *The Scientific Estate*. Cambridge, MA: Harvard University Press.

Rashad, R., ed. 1988. *Sciences à l'époque de la Révolution française*. Paris: Blanchard.

Rausch, Jean-Marie. 1987. *Le laminoir et la puce: La troisième révolution industrielle*. Paris: Lattès.

Rémond, René, Aline Coutrot, and Isabel Boussard. 1982. *Quarante ans de cabinets ministériels*. Paris: Presses de la fondation nationale des sciences politiques.

Ridley, Frederick F., and Jean Blondel. 1964. *Public Administration in France*. London: Routledge and Kegan Paul.

Rouban, Luc. 1987. *L'Etat et la science*. Paris: Editions du Centre National de Recherche Scientifique.

Rousseau, Pierre. 1974. *Survol de la science française contemporaine, 1939–1974*. Paris: Fayard.

Salomon, Jean-Jacques. 1986. *Le gaulois, le cow-boy, et le samouraï: La politique française de la technologie*. Paris: Economica.

Schatzman, Evry Léon. 1989. *La science menacée*. Paris: Odile Jacob.

Scheinman, Lawrence. 1965. *Atomic Energy Policy in France under the Fourth Republic*. Princeton: Princeton University Press.

Spiegel-Rosing, Ina, and Derek John de Solla Price. 1977. *Science, Technology and Society*. London: Sage.

Smith, Cecil O., Jr. 1990. "The Longest Run: Public Engineers and Planning in France." *American Historical Review* 95: 657–92.

Suleiman, Ezra. 1974. *Politics, Power, and the Bureaucracy in France*. Princeton: Princeton University Press.

Suleiman, Ezra. 1978. *Elites in French Society*. Princeton: Princeton University Press.

Weart, Spencer. 1980. *La grande aventure des atomistes français*. Paris: Fayard.

Wood, Robert C. 1964. "Scientists and Politics: The Rise of an Apolitical Elite." In *Scientists and National Policymaking*, ed. Robert Gilpin and Christopher Wright. New York: Columbia University Press.

Japan: The Political Economy of Japanese Science: Nakasone, Physicists, and the State

Morris F. Low

This chapter examines how the organization of science in Japan reflects political and economic structures, which in turn are shaped by Japan's involvement in the international context. It firstly provides a historical survey of postwar Japan, emphasizing how various phases signalled the emergence of conceptions of the political economy of science relevant to the study of state-scientists interaction. The second part focuses on former prime minister Yasuhiro Nakasone, a major player in translating those conceptions into action. Physicists responded to initiatives taken by him, amid an environment of external conflicts and a push for international economic competitiveness. We look at two areas in particular: science policy and more specifically atomic energy. In the third and final part, we reflect on what this tells us about state-scientists relations. We will find that physicists have not acted as one in Japan, some choosing to converge closer to the interests and aims of the state, while others have located themselves on the periphery in positions of conflict.

Japan's Political Economy

A detailed examination of the role of physicists shows how ideal conceptions of democratic participation in policy-making differ from practical 'democratic' realities. One important component of a democracy are interest groups which serve to link citizens to government. Scientists have been highly organized as an interest group and warrant examination. Japanese physicists were especially influential in policy-making during the postwar period. While Japanese science policy has been dominated by the power elite of big business, government (the Liberal Democratic Party), and bureaucracy, Japanese public policy has also been the result of 'democratic' bargaining among a number of diverse groups.

Such questions have been traditionally addressed in terms of two alternate models which can be described as *elitist* and *pluralist*.[1] The former tends to emphasize the concentration of power in a particular elite group. In the case of Japan it is argued that a tripartite power elite consisting of the Liberal Democratic Party (L.D.P.), bureaucracy, and big business has controlled decision making on major issues of industrial policy since the end of the Allied Occupation. The pluralist perspective, however, argues that public policies are more the result of negotiation or bargaining among diverse groups, including citizens, unions, and academics.[2] The pluralist model points to conflicting opinion within groups, to factions within the L.D.P., and to inter-ministry factionalism within the government bureaucracy.[3]

The pluralist perspective argues that the elite model neglects other groups, which it portrays as being unable to influence policy decisions. The pluralist perspective offers a more complex view of policy-making. I would argue, however, that the two views are not incompatible, and like the relationship between science and technology, represent "two ends of a continuum,"[4] in which science policy flows from business leaders, bureaucrats, and politicians after consultation with citizens and scientists. In an attempt to find evidence to support such a view, this chapter examines how physicists influenced the policy-making process in science and science-based technology by looking at the interaction between physicists and other groups (e.g., big business, bureaucracy, L.D.P.) involved in science policy-making. The case of the policy-making activities of Yasuhiro Nakasone provides a useful window to the political economy of science in Japan.

The sacrifice of life at Hiroshima marked the beginning of the nuclear age throughout the world and a new phase in the nation's political economy. Many physicists were somehow touched by the atomic bomb. The prominent physicists Yoshio Nishina, Shôichi Sakata, Mituo Taketani, and Hideki Yukawa were consultants to Japan's wartime atomic bomb project. The bomb

1. See for example the discussion in Brian W. Hogwood and Lewis A. Gunn, *Policy Analysis for the Real World* (Oxford: Oxford University Press, 1984), 71.

2. T. J. Pempel has suggested that Japanese policy-making is extremely heterogenous, and that there is no one single policy-making process in Japan. No one group plays an absolute role in decision making. He argues that the diversity in policy-making defies existing generalizations which often characterize it in terms of factionalism and cultural traits. The wide differences in opinion on policy issues amongst interest groups in Japan make the notion of a harmonious 'Japan, Inc.' difficult to sustain. See T. J. Pempel, ed., *Policymaking in Contemporary Japan* (Ithaca: Cornell University Press, 1977), 308–10.

3. Haruhiro Fukui, "Studies in Policymaking: A Review of the Literature," in Pempel, *Policymaking*, 22–59, esp. 22.

4. Roya Akhavan-Majid, "Telecommunications Policy-making in Japan, 1970–1987: A Case Study in Japanese Policy-making Structures and Process," Ph.D. diss., University of Minnesota, 1988, 7.

would prove fatal for Japanese physicists who were members of survey groups. Nishina entered Hiroshima on 8 August. The Tokyo University professor Ryôkichi Sagane also joined a group to inspect Hiroshima, as well as leading a group to Nagasaki in which Satio Hayakawa was a member. Nishina, Sagane, and Hayakawa would all die relatively early, at the ages of 61, 63, and 68, respectively.

While the atomic bomb may have cost them their lives, it was ironically the bomb which invested their lives with a greater authority. It placed physicists who understood the atom in the position of helping to harness its power for economic, scientific, and political ends. A generation of young Japanese physicists was suddenly catapulted into the public arena and transformed into a highly influential intellectual group. The political activities of this progressive group of scientists were, for a short period after the war, encouraged by the Occupation Forces but nuclear research was effectively ruled out. The American physicists Harry Kelly, Ernest Lawrence and I. I. Rabi figured prominently in the attempt to remake Japan in the image of U.S. democracy, and Japanese science in the mold of Berkeley physics.

The Allied Occupation of Japan

The gaining of a political voice added to the call for democracy and science in service to the nation immediately after World War II. While Headquarters of the Supreme Commander for the Allied Powers (S.C.A.P.) introduced a 'democratic' element by incorporating scientific and scholarly expertise into the reorganized policy-making structure, prime ministers would be under no compulsion to heed the decisions made by these experts. Nishina and Sagane felt that science could be used for practical ends such as medical research (e.g., radio-isotopes and the production of penicillin) and nuclear power. For them, physics had no true path of its own; rather, it lent itself to increasing opportunities for integration into industry. Left-wing physicists such as Sakata and Taketani viewed the role of science in terms of Marxism being able to change society. More apolitical physicists, such as Yukawa and Sin-itirô Tomonaga, felt that science could contribute through adding to the bank of basic knowledge. And Hayakawa realized that research infrastructure and keeping abreast of international development was necessary to achieve this.

The Occupation authorities encouraged the Japanese to form democratic and representative institutions in an attempt to restructure the political system. A new constitution was written, both houses of parliament became subject to popular elections, and parliament became the highest organ of state power. Leftists established the Japan Socialist Party and Japan Communist Party in November, 1945. The conservatives organized themselves as well, with the creation of the Japan Liberal Party and Japan Progressive Party that month. The Japan Cooperation Party was established the following month.[5]

By late 1947, U.S. leaders increasingly saw Japan, rather than the Chinese Nationalist Government, as the means by which to contain the spread of Communism in Asia.[6] It can be shown that President Herbert Hoover urged Truman to consider both Japan and Germany as potential allies against the Soviet Union from as early as 1945. Hoover favored the reconstruction of Japan as a bulwark against Communism from the very beginning of the Occupation;[7] Japan should be dealt with kindly and allowed to retain its colonies of Korea and Taiwan. Major General Charles A. Willoughby, who headed the Intelligence Section G-II of S.C.A.P., shared Hoover's view that the fight against Communism should take precedence over the democratization of Japan. As soon as three days after the Occupation had begun, Willoughby began recruiting high-ranking Japanese officers for his organization. One group of Japanese officers was involved in activities against the Japan Communist Party. Willoughby also was involved in arranging immunity for Japanese scientists involved in chemical and biological warfare research.[8] It appears that both Prime Minister Shigeru Yoshida and General Douglas MacArthur were unaware of many of these activities.

Many physicists had hopes that the Science Council of Japan (J.S.C.) would become the central organization for science and technology, but in the science and engineering boom which followed in the next ten years they were no match for state-led objectives and strategies. The shortage of physicists experienced by Japan during World War II made the training of scientists and engineers for industry a leading priority after the war. By 1950, this increasing pool of scientific and technical personnel generated by the universities was no longer confined to university or government laboratories, but was also directed into the private-sector laboratories of industry.

There was a greater emphasis placed on applied science and technology than upon pure science. This placed a major tension upon the pure state-science relationship, serving to differentiate science-based, basic research in the universities from research that occurred in the private sector and was more closely aligned with economic goals.

The importance placed on economic growth and the Occupation experi-

---

5. Nathaniel B. Thayer, *How the Conservatives Rule Japan* (Princeton: Princeton University Press, 1969, 1973), 7.

6. John Welfield, *An Empire in Eclipse: Japan in the Postwar American Alliance System: A Study in the Interaction of Domestic Politics and Foreign Policy* (London: Athlone Press, 1988), 26–27.

7. Michael Schaller, *The American Occupation of Japan: The Origins of the Cold War in Asia* (New York: Oxford University Press, 1985), 9–19, 92–94.

8. Welfield, *An Empire in Eclipse*, 66–67.

ence have led to a heightened sensitivity to forces controlling the international environment, policy often being shaped by reactions to external events. International pressure (*gaiatsu*) has thus been a major agent of domestic transformation, especially when strong market forces and Japanese interests abroad are supported by domestic interest groups in Japan. With the increasing globalization of R&D, gaiatsu will be of even greater importance in determining the direction of science in the future.[9] The concept provides a useful key to understanding how the international context influences the internal structures of the political-economic system, and the state-science relationship broadly discussed by Solingen in chapter 1.

Partners in Big Science

In the rush to harness atomic energy after the Allied Occupation, the atom brought together politicians and physicists such as Sagane and Yukawa. Even the left-wing Taketani encouraged Japan to pursue the promise of nuclear power. The ensuing struggle to influence the organization of science was particularly exacerbated in the context of the development of nuclear power, the aerospace industry, accelerator-based big science, and the government-industrial complex. The hitherto appearance of widespread consensus and lack of conflict in Japanese policy-making has often been achieved by the use of public authority to privatize conflict rather than to socialize societal conflict.[10] Privatization of conflict via government-aligned policy-making structures has reduced the number of participants, benefited insiders rather than outsiders, and limited public scrutiny of decision-making processes. Outsiders (such as Sakata and Taketani) turned to what Titus calls socialized means of pressure—newspapers, protests, and petitions to convince policymakers of the validity of their positions.[11]

Reciprocal Consent

A coherent science policy is difficult to find in Japan. It can be said that self-sufficiency has been used as an argument or replacement for policy, as has the

9. Kenneth B. Pyle, "The Burden of Japanese History and the Politics of Burden Sharing," in *Sharing World Leadership?: A New Era for America and Japan* ed. John H. Makin and Donald C. Hellmann (Washington, D.C.: American Enterprise Institute for Public Policy Research, 1989), 41–77, esp. 50–51; Kent E. Calder, *Crisis and Compensation: Public Policy and Political Stability in Japan, 1949–1986* (Princeton, N.J.: Princeton University Press, 1988), 117–26.

10. This has been discussed in Chalmers Johnson, *M.I.T.I. and the Japanese Miracle: The Growth of Industrial Policy, 1925–1975* (Stanford: Stanford University Press, 1982), 8.

11. David Anson Titus, *Palace and Politics in Prewar Japan* (New York: Columbia University Press, 1974), 312, 332.

policy of working hard and reacting swiftly to changes in global conditions such as the oil shock of 1973. The latter was a prime example of the impact of international economic developments on domestic politics, a form of gaiatsu referred to earlier.

Of all the advanced industrial nations Japan has the smallest public sector and the lowest ratio of public R&D expenditure to GNP, and yet the government and the bureaucracy can wield a great deal of influence. Policy-making has been regarded as a joint operation in which both government and industry have aimed at the same target of rapid industrial growth. Public policies in science and technology have been highly compatible with the aims of private firms, resulting in the production of market-oriented know-how, as Solingen has pointed out, not unlike Germany and Sweden.

R&D in Japan has been mainly funded by private industry rather than government. This ratio is the highest of all industrial nations. If we include government grants-in-aid to universities, the government share in 1983 was only 22.2 percent (dropping to 17.9 percent in 1990). Of this small sum, 40 percent was directed to agriculture and another 40 percent to energy research. Total government R&D funding for the United States over the same time was more than double, with 46 percent. The Japanese private R&D funding share is still higher than that in the United States even if we exclude military R&D. Using figures for nondefense R&D, Japan's share of GNP devoted to R&D has exceeded that of the U.S. since the early 1960s. In 1985, R&D expenditure rose to 2.77 percent, higher than the U.S. GNP share of 2.4 percent.[12]

Such factors have resulted in an emphasis on engineering rather than basic science in Japanese universities. Of all natural science and engineering students, only 12 percent hold first degrees in science, whereas the same figure is around 56 percent for the United States.[13] Despite the emphasis on engineering, those Japanese scientists who are employed in the private sector are able to focus on the development of commercial products rather than defense research. Once scientists have entered company or university laboratories, there is a lack of mobility and their relations with the state are often defined by their workplace. In order to promote greater dissemination of

---

12. David C. Mowery and Nathan Rosenberg, *Technology and the Pursuit of Economic Growth* (Cambridge: Cambridge University Press, 1989), 221; Sully Taylor and Kozo Yamamura, "Japan's Technological Capabilities and Its Future: Overview and Assessments," in *Technological Competition and Interdependence: The Search for Policy in the United States, West Germany, and Japan* ed. Gunter Heiduk and Kozo Yamamura, (Seattle: University of Washington Press/Tokyo: University of Tokyo Press, 1990), 25–63, esp. 33. Also, Japanese Government, *Kagaku gijutsu hakusho (Heisei yonnen ban) (Science and Technology White Paper)* (Tokyo: Ministry of Finance Printery, 1992), 167–68.

13. Statistics (c. 1985) cited in Taylor and Yamamura, "Japan's Technological Capabilities and Its Future," esp. 28.

know-how, the Ministry of International Trade and Industry (M.I.T.I.) has created national research projects aimed at producing commercially viable products and designed to overcome barriers among scientists.

Richard J. Samuels has recently written of the political interdependence of states and markets, what he calls the politics of "reciprocal consent" where public officials and market players accommodate each other. Relations between producers, consumers, and the state affect the ability of actors to pursue certain strategies. As a result of reciprocal consent, the triumvirate of bureaucracy, big business, and the L.D.P. have dominated policy-making in Japan. Through bureaucratic, state control of supposedly independent advisory committees, and through ordinance power, bureaucrats and ex-bureaucrats, along with the ruling L.D.P., have been able to exercise a great deal of control over the policy-making process in postwar Japan. Centralization of state power in the bureaucracy enables better coordination of the politics of reciprocal consent. Ex-bureaucrats often join the Party upon retirement, and have constituted at least 30 percent of L.D.P. membership in the Diet since the formation of the party.[14] In the period from 1955 to 1977, 50 percent of the prime ministers were ex-bureaucrats.[15] This has all served to blur the divisions between the bureaucracy and the L.D.P.

The Power of Bureaucrats

Technocrats can often monopolize policy implementation. Even with advisory bodies, control of the investigation process, the synthesis of discussion, and final compilation of the report is placed in the hands of bureaucrats. This is not to say that politicians are not highly influential. Muramatsu and Krauss suggest that politicians have influenced policy-making more than is commonly thought.[16] This has led many to conclude that politicians have reigned and bureaucrats ruled.[17] The L.D.P. exercises a great deal of influence over promotions within the bureaucracy.[18] But the reverse is true too. Many ex-bureaucrats occupy influential positions in L.D.P. policy-making bodies as well as holding cabinet positions. In November, 1963, it was estimated that

---

14. Michio Muramatsu and Ellis S. Krauss, "Bureaucrats and Politicians in Policymaking: The Case of Japan," *The American Political Science Review* 78 (1984): 126–46, esp. 128–29.

15. Roger Benjamin and Kan Ori, *Tradition and Change in Postindustrial Japan: The Role of the Political Parties* (New York: Praeger, 1981), 40–41.

16. Muramatsu and Krauss, "Bureaucrats and Politicians," 126–46.

17. Johnson, *M.I.T.I. and the Japanese Miracle,* 35, 317; Byung Chul Koh, *Japan's Administrative Elite* (Berkeley: University of California Press, 1989), 256–57.

18. T. J. Pempel, "The Bureaucratization of Policymaking in Postwar Japan," *Midwest Journal of Political Science* (later renamed *American Journal of Political Science*) 18 (1974): 647–64.

ex-bureaucrats comprised 26 percent of the L.D.P. members in the House of Representatives.[19]

Japan's postwar prime ministers have been close to the bureaucracy. During the years between Japan's defeat and 1978, seven of the thirteen prime ministers have had bureaucratic backgrounds, perhaps the key to their longevity as premiers. These include Shigeru Yoshida, Hayato Ikeda, and Eisaku Satô. The "magnificent seven" retained the prime ministership for a total of twenty-four of those thirty-three years. Six of the seven were graduates of Tokyo University.[20] Politicians with bureaucratic experience understand the policy-making process and have the power to realize plans which physicists have formulated.

Pluralistic Structure of Big Science

The development of atomic energy and space research show the existence of a pluralistic structure, the result of trade-offs between scientists, industrialists, politicians, and bureaucrats. The trade-offs occurred against the backdrop of postwar economic growth and a belief in the need for rapid modernization, which was characterized by: high returns from importing technology in terms of exports and productivity; extensive government controls over importation of technology; and clever use of management, investment, and domestic R&D to capitalize upon imported technology. This has occurred against the backdrop of concerns about Japan's national security.

Japan's Rearmament

Japan's rearmament can be considered to have been triggered by Truman's decision to intervene in the Korean civil war. In April, 1950, John Foster Dulles was appointed top adviser to the Secretary of State, Dean Acheson, for Asian matters, and especially the peace treaty with Japan.[21] Dulles arrived in Tokyo in June, 1950, urging Prime Minister Yoshida to greatly expand its armed forces, in addition to allowing U.S. bases to remain in Japan.[22] The

---

19. Shigeo Misawa, "An Outline of the Policy-making Process in Japan," in *Japanese Politics—An Inside View: Readings from Japan* ed. and trans. Hiroshi Itoh (Ithaca: Cornell University Press, 1973), 12–48, esp. 22.

20. Hyôe Murakami, "The Making of a Prime Minister," in *Politics and Economics in Contemporary Japan* ed. Hyôe Murakami and Johannes Hirschmeier (Tokyo: Kodansha International, 1983), 3–20, esp. 3–4.

21. Michael Schaller, *Douglas MacArthur: The Far Eastern General* (New York: Oxford University Press, 1989, 1990).

22. Meirion Harries and Susie Harries, *Sheathing the Sword: The Demilitarisation of Japan* (London: Hamish Hamilton, 1987), 235–36.

Yoshida government was ordered by MacArthur in July, 1950 to establish a Police Reserve Force consisting of 75,000 men. The Maritime Safety Force would also be increased by 8,500 men.[23] In August 1950, the Police Reserve Force was formed and this later became the Peace Preservation Agency on 1 August 1952. That very same day, the Peace Preservation Agency Technical Research Institute was established and with it the recommencement of what some commentators view as military research.

Dulles returned for a further round of talks in January and February, 1951. In the summer of 1952, the Federation of Economic Organizations (F.E.O. or *Keidanren*) established the Defense Production Committee. The Arms Production Cooperative Association was started and this later became the Japan Arms Industries Association (*Nihon Heiki Kôgyôkai*). Its aims were to conduct studies and research into the technical side of weapons.

Despite the demilitarization of Japan during the Occupation, it was the United States that encouraged Japan to increase arms production in the 1950s, in order to support American forces in Korea during the Korean War, and later in Vietnam.[24] Arms production is thus considered as having played a crucial role in Japan's postwar economic growth. For example, in the period 1951–60, U.S. military procurements totalled $U.S. 6 billion, that is an average of $U.S. 600 million each year. In 1958–59, such revenue was sufficient to cover the cost of 14 percent of Japanese imports.[25] War was helping to feed the Japanese.

As in other countries, security considerations prompted attempts to establish a self-sufficient arms industry in Japan as well. However, military-related R&D has never approached the levels found in the United States or former Soviet Union discussed by Solingen in chapter 1. In April, 1953, the Arms Production Law was promulgated and in September the Peace Preservation Agency announced its five-year plan. Despite the 'demilitarization' of Japan during the Occupation, 80–90 percent of Japan's armament production capacity was intact at the beginning of the Korean War.[26] In 1952, a joint business and military Committee for Defense Production was established to institutionalize the links between the military and industry. It was thus to be

23. Welfield, *An Empire in Eclipse*, 72.

24. Malcolm McIntosh, *Japan Re-armed* (London: Frances Pinter, 1986), 50–51.

25. Jon Halliday and Gavan McCormack, *Japanese Imperialism Today: 'Co-Prosperity in Greater East Asia'* (New York: Monthly Review Press, 1973), 107–8. In the period 1965–66, Japan's G.N.P. rose 2.7 percent but the following year, thanks to the escalation of the Vietnam War, the G.N.P. rose 7.5 percent. In that year of 1966–67, military contracts were worth $505 million, and another $1.2 billion worth of contracts for delivery of goods to the United States, Thailand, South Korea, Taiwan, and the Philippines were war related. It can thus be argued that Japan's economy was shaped by war demand.

26. Halliday and McCormack, *Japanese Imperialism Today*, 108.

expected that in early 1953, Japanese business organized for rearmament. In October, 1953 the Arms Production Council was started within M.I.T.I., under which eight subcommittees were created the following March to look into different technical aspects. On 14 November 1953, Kimura, the director of the Peace Preservation Agency, stated that he had hopes of expanding the Technical Research Institute and including in its research activities radio-guided missiles and atomic energy. It seems that, when appropriate, the development of atomic energy for peaceful purposes and its development for bombs were interchangeable.[27] At the end of 1953, a Guided Missile Forum was jointly established by the Arms Production Committee, Arms Industries Association, and the Aeronautic Industries Association. In March, 1954, the Mutual Security Agreement was signed, under which Japan was to receive weapons and other forms of aid. In May, the Defence Secrets Protection Law was established, followed by the creation in June of the Defense Council.[28] In 1954, however, arms production still accounted for a low 3.9 percent of machine industry production.

In 1954 Japanese business lobbied for research into guided missiles, and the following year orders were obtained for the production of jet fighters.[29] In July, 1954, a Guided Missile Research Committee was established within the Peace Preservation Agency. From September, that agency became the Defense Agency.[30] It was thus only two years after the end of the Occupation that Japan had acquired a new army, navy, and air force. This was courtesy of legislation such as the Self Defense Forces Law of June, 1954, and the Law Concerning the Structure of the National Defense Council in July, 1956.[31]

Nuclear Power

Despite the uncertainties surrounding nuclear technology and opposition from citizen groups, the development of nuclear power continues unabated in Japan. This is similar to the situation in countries such as the former U.S.S.R., France, and Germany, but is in strong contrast to the situation in the United States, where pressure from the community and the sheer economics of the matter has meant that it is sometimes better to close down a reactor before

---

27. *Asahi shinbun,* 15 November 1953. Cited in Tetu Hirosige, *Sengo Nihon no kagaku undô (Postwar Japan Science Movement)* (Tokyo: Chûô Kôronsha, 1960), 90.

28. Hirosige, *Postwar Japan Science Movement,* 89.

29. Halliday and McCormack, *Japanese Imperialism Today,* 107–8.

30. Mitutomo Yuasa, *Kagakushi (The History of Science)* (Tokyo: Tôyô Keizai Shinpôsha, 1961), 318: Hirosige, *Postwar Japan Science Movement,* 90. The Research Institute was renamed the Defence Agency Technical Research Headquarters in 1959. By 1960 the budget for the institute had become over ¥2.1 billion ($U.S. 5.83 million).

31. Welfield, *An Empire in Eclipse,* 61.

even starting. A feature of the nuclear power program in Japan is the fact that responsibility for the electricity supply, ownership of nuclear fuel, and commercial power reactors is concentrated in the hands of private industry. Some scientists and members of the public have long felt that this tends to make questions of public safety secondary to questions of self-interest.

The determination of the government and industry to promote the commercialization of nuclear power, in the face of opposition from some scientists and more recently the general public, has been a feature of the development of the nuclear power industry since its very beginning. The example of nuclear power shows that Japan's rapid postwar development was, more than anything else, due to the close collaboration between organized business, government, and the scientific establishment.

In 1946, the Atomic Energy Act was passed by the U.S. Congress. This, in effect, created the Atomic Energy Commission (A.E.C.) and placed severe restrictions on the dissemination of nuclear research. The United States had hopes after the war of forming an alliance with a China under Chiang Kai-shek. Japan was to be demilitarized, the leaders of its war effort purged, and Japan would then become the 'Switzerland of the Far East'. The role of Japan in U.S. strategy changed, however, with the rise of the cold war, the Communist takeover in China and the beginning of hostilities in Korea in June, 1950. The change in the political situation coincided with what many commentators consider as the rebuilding and strengthening of Japan's military establishment. Japan became a bastion against Communism and the 'workshop of Asia': a source of supplies and services for the Korean War. The end of the Allied Occupation in 1952 and the signing of the U.S.-Japan Security Treaty was a coming-of-age for Japan as the chief East Asian ally of the U.S. The Korean War and cold war politics effectively enabled the conservatives to come back into favor in Japan and for Japanese industry to be propelled by war demand, i.e., the requirements of the U.S. Armed Forces in Japan and military requirements in the Korean War.

The development of nuclear power in Japan bears similarities to that in the Federal Republic of Germany. Due to the terms of the end of World War II and the cold war politics that followed, Japan and the F.R.G. initiated programs a decade after that of the U.S.S.R., United States, and Canada. Furthermore, when programs did commence in Japan and the F.R.G., most of the technology was imported and materials such as enriched uranium were obtained via the U.S. Atoms for Peace Plan. As a result, the prohibitions placed on the development of aerospace technology and nuclear power by the Allied Powers immediately after the end of the war effectively provided the United States with a ready customer in both of these areas, for the prohibitions on research related to nuclear energy did not prevent growing interest on the part of physicists and industrialists in the scientific and commercial possibilities offered by this new energy source. Conscious or not, Occupation policy became linked with the creation of U.S. markets in Japan, an early form of gaiatsu if ever there was one.

While the need for basic research was urged by physicists, the government's Science and Technology Agency was intent on pushing a policy with the aim of developing facilities for harnessing nuclear power. This line of policy resulted in Japan later becoming the fourth largest user of nuclear power after the United States, France, and U.S.S.R., but at the cost of interrupting the steady, well-balanced progress in the basic sciences that the scientific community had hoped for in the postwar period.

The physicist Mituo Taketani proposed three principles for the peaceful use of atomic energy, calling for public disclosure of atomic research, democratic management of research, and research autonomy. These had a major impact on the formulation of the Basic Atomic Energy Law passed by the Diet in 1955, which forbid use of atomic energy for military purposes. The Three Non-Nuclear Principles later declared by Prime Minister Eisaku Satô in 1967, amounted to a restatement of these three principles and the Basic Atomic Energy Law.

The "1955 System"

In November, 1955, with the formation of the third Hatoyama government, the business world was convinced that the conservative parties should join forces to form a strong, stable government. The Liberal Party (which had long supported Shigeru Yoshida's policies) and the Japan Democratic Party amalgamated to form the Liberal Democratic Party, a party that has since been characterized by its emphasis on economic development, a close relationship with big business, and cooperation with the United States.[32] This has helped to shape a policy-making structure which has been described as the "1955 system," a system, lasting until the early 1970s, that saw close linkages between business and the L.D.P., electoral invincibility, clear policy directions in foreign policy and economic policy, and relative unity among the conservatives.[33] It can be argued that the Japanese parliament often merely ratifies decisions made elsewhere by big business, the bureaucracy, and the L.D.P. This has been exacerbated by the fact that the L.D.P. has continually been returned to office since its formation in 1955. Whether the 1955 system can survive beyond the early 1990s is a topic of much debate.

32. Janet E. Hunter, *Concise Dictionary of Modern Japanese History* (Berkeley: University of California Press, 1984), 113–14.

33. T. J. Pempel, "The Unbundling of 'Japan, Inc.' The Changing Dynamics of Japanese Policy Formation," *Journal of Japanese Studies* 13, no. 2 (Summer 1987): 271–306.

The Japan Atomic Energy Commission

Formal government authority for nuclear power came to rest in the Prime Minister's Office in the form of the Japan Atomic Energy Commission (J.A.E.C.), which was established under a special law on 19 December 1955. The first meeting of the J.A.E.C. was held on 4 January 1956 and included physicist-members Hideki Yukawa and Yoshio Fujioka.[34] Its activities reflect the 1955 system at work.

Growth in new government bodies coincided with the organizational needs of atomic energy and the political desire to avoid troublesome physicists. Special needs of the nation gave rise to the creation of a host of bodies such as the J.A.E.C., Japan Atomic Energy Research Institute and the National Space Development Agency, all which facilitated the importation of foreign technology and enabled Japan to bypass the need for much costly and time-consuming R&D.

The J.A.E.C. began on 1 January 1956, only to be startled by a statement on 5 January by the chairman, Matsutarô Shôriki, the day after the first meeting of the Commission. (Shôriki was president of the *Yomiuri shinbun* newspaper company and later Cabinet minister.) Shôriki announced a target of building a power reactor within five years. At the time, there was a willingness among physicists to develop nuclear power independently, and only after a technology base had been created. Yukawa, a J.A.E.C. member, threatened to resign. The Commission attempted to defuse the situation by releasing a more modest statement which watered down Shôriki's words. It was announced that the Commission was determined to succeed in actually generating atomic energy within five years.[35] From the time of Shôriki's controversial statement, Yukawa felt that taking on membership of the J.A.E.C. had been a mistake. Yukawa was persuaded not to resign. He was told that if he were to do so, it would cause a great controversy, journalists would make much fuss, and even the Cabinet could be put into danger. Yukawa gave into such representations, but in March of the following year he resigned, citing ill health as the reason.[36]

In May, 1956, the Science and Technology Agency was established, headed by none other than Shôriki, who was also a state minister and chairman of the Atomic Energy Commission. Shôriki had also been one of the

---

34. Chitoshi Yanaga, *Big Business in Japanese Politics* (New Haven: Yale University Press, 1968), 193–94.

35. Japan Atomic Industrial Forum, *Nihon no genshiryoku: 15 nen no ayumi, jyô (Atomic Energy in Japan: A 15 Year History, Part 1)* (Tokyo: Japan Atomic Industrial Forum, 1971), 58–59.

36. Japan Atomic Industrial Forum, "Atomic Energy," 61–62.

principals behind the creation of the Japan Atomic Industrial Forum in March, 1956.[37]

Industrial Policy

From 1952 to 1960, extensive government controls were administered by M.I.T.I. The government aimed at making Japan more self-sufficient in materials such as chemicals and iron and steel. In the period 1960–65, imports of technology became more directed toward consumer goods with export markets in mind. During 1966–72, the Japanese consolidated this base by using imports of technology increasingly to improve technology that had been previously imported. To the physicists' dismay, Japan did not depend on its own pioneering R&D in order to prosper.[38] Even in nuclear power this was the case, with a strategy of importing technology being emphasized, along with some domestic R&D. Through M.I.T.I., the state was able to act as a gatekeeper of science and technology, effectively deciding what could enter or leave Japan and mediating between domestic structures and the international context.[39]

Response of Scientists

Scientists have attempted to counter this over-reliance on foreign technology by stressing the importance of scientific activity and some of the supposed values associated with it in the West, best met within the confines of academic institutions. Efforts were made by physicists after the war to maintain the group collaboration which had been a characteristic of prewar and wartime research in Japanese particle physics. This was despite the inevitable splitting of physicists into differing specializations and factions.

The Research Institute for Fundamental Physics (R.I.F.P.) at Kyoto University strived to bring physicists together within the context of a university but via the concept of interuniversity research institutes. Leading physicists devoted much time and energy to setting up research institutes throughout Japan which effectively provided them with a power base from which to negotiate power. The Institutes were firstly attached to national universities, but their proliferation and the sheer scale and demands of big science proved

37. Yanaga, *Big Business,* 195–97.

38. Merton J. Peck, with Shûji Tamura, "Technology," in *Asia's New Giant: How the Japanese Economy Works,* ed. Hugh Patrick and Henry Rosovsky (Washington, D.C.: Brookings Institution, 1976), 525–85, esp. 527, 535, 558.

39. Masaru Tamamoto, "Japan's Search for a World Role," in *Power, Economics and Security: The United States and Japan in Focus* ed. Henry Bienen (Boulder, Colo.: Westview Press, 1992), 226–53, esp. 233–34.

too much for host institutions, such as the University of Tokyo, to handle. The K.E.K. High Energy Physics Laboratory and the Institute of Space and Astronautical Science symbolize the change from university-affiliated institutions to autonomous, truly national, common-use research institutes. Group collaborations took on new meaning with the emergence of big, group-oriented science centered around an accelerator and other facilities.

Lobbying for Funds

To support the lobbying for the establishment of research institutes and funding of large projects, individual specializations within physics had to organize themselves across university boundaries to enable their voices to be heard. But R&D funds were finite and the distribution of funds to those supposedly most deserving conflicted fundamentally with the democratic notion of equal distribution of funding within a community. These problems would cause dissension among physicists and result in greatly delaying the construction of Japan's large-scale accelerator.[40]

There were rumblings of concern emanating from ministries and the L.D.P. that the control of large budgets should not be placed in the hands of left-leaning physicists. Such huge funds had to be 'properly' managed—hence the debate over physicist-led management of the Institute for Nuclear Study (I.N.S.) at Tokyo University, and the similar controversy which stalled the establishment of the K.E.K. High Energy Physics Laboratory at Tsukuba. It was inevitable that the government would wish to manage such facilities in the 'national interest.'[41] But when Japanese physicists tried to invoke the rhetoric of national interest when arguing for expensive accelerator projects, they met with resistance. Such machines were not considered part of the atomic energy program.

Two-Tiered Policy-Making

These tensions gave rise to the emergence of two tiers for managing governmental research. Advisory bodies such as the Council for Science and Technology and the Japan Science Council, and organizations like the Science and Technology Agency (S.T.A.) tend to provide central policy coordination and make up one group. But there are also ministries which have their own

40. See the discussion of elitism vs. democracy in Catherine Lee Westfall, "The First 'Truly National Laboratory': The Birth of Fermilab," 2 vols., Ph.D. diss., Michigan State University, 1988, esp. 25.

41. See the situation in Great Britain described by Tom Wilkie, *British Science and Politics since 1945* (Oxford: Basil Blackwell, 1991), esp. 124–25.

advisory and coordinating bodies. It should be noted, however, that the S.T.A. and M.I.T.I. account for over 70 percent of the government R&D commitment[42] and the industrial sector funds approximately 80 percent of the country's total research effort.

The fighting for 'territorial' rights among S.T.A., M.I.T.I., and the Ministry of Education has resulted in overlap in management of R&D in at least two cases involving both imported technology and local R&D: nuclear power and the aerospace industry. In the former, the government and industry were determined to promote the commercialization of nuclear power, in the face of opposition from some physicists who advocated a more gradual approach. This manifested itself in the creation of the Japan Atomic Energy Commission (J.A.E.C.), which was established under a special law on 19 December 1955. In 1956, the Japan Atomic Energy Research Institute (J.A.E.R.I.) was created to conduct R&D. The S.T.A. was also launched that year to manage nuclear R&D and provide support for the Commission, along with the Japan Atomic Industrial Forum (J.A.I.F.). The Japan Atomic Fuel Corporation (J.A.F.C.) followed soon after in 1957. M.I.T.I. took on the responsibility for actual nuclear power generation and in 1959, the Science and Technology Council was created as a government review body which would act as a buffer between the government and sometimes hostile J.S.C.

What thus resulted was a pluralistic structure for the promotion of nuclear power consisting of two groups: (1) M.I.T.I., the atomic power industry and electric power companies; and (2) the S.T.A. and public corporations for research. Research became the domain of the second group, but light-water reactors (L.W.R.s) became the area of focus for the first group. This has made for a strong contrast between the policies of M.I.T.I. and the S.T.A. The former has concentrated on nurturing infant industries and promoting policy for profit, with the interests of business in mind whereas the S.T.A. has tended to participate in national projects at the forefront of technology. This can be interpreted as the result of private industry attempts to socialize the risks of nuclear development.[43]

Space R&D

A characteristic of this pluralistic structure in the management of R&D can be seen particularly clearly if we compare the development of nuclear power to space development. The latter is organized under the auspices of S.T.A., on the one hand, and the Ministry of Education, Science, and Culture (*Mon-*

42. J. Ronayne, *Science in Government* (Melbourne: Edward Arnold, 1984), 210, 216.

43. See Richard J. Samuels, *The Business of the Big State: Energy Markets in Comparative and Historical Perspective* (Ithaca: Cornell University Press, 1987), 234.

bushô) on the other. For a long period, the University of Tokyo's Institute of Industrial Sciences and its successor, the Institute of Space and Aeronautical Science, were the Japanese space program. In 1962, however, the Technical Research and Development Institute of the Japan Self-Defense Agency established the Niijima rocket and missile test center. Several years later, the Science and Technology Agency proceeded with its own space program, which differed substantially from that of University of Tokyo. The Agency established the National Space Development Agency of Japan (N.A.S.D.A.), with the initial aim of launching a satellite. In contrast to the Tokyo University group, which insisted on developing technology domestically, N.A.S.D.A. adopted U.S. technology.

There are many government agencies, research institutes, industries, and committees involved in space research but the organization and coordination of research activities has been, as in the case for nuclear power, sadly lacking. The three most important government bodies for space research each have their own facilities and programs: the Institute of Space and Astronautical (formerly Aeronautical) Science is now a national research institute and has the Kagoshima Space Center; the Self-Defense Agency has the Niijima Test Center; and there is also the National Space Development Center on Tanegashima. For better or for worse, physicists have influenced the emergence of this pluralistic structure which is evident in a number of scientific fields which are of both academic and commercial interest. Such domestic structural constraints fragment state authority and render Japan's global role a reactive one, responding to gaiatsu when the need arises.[44] Political instability in the 1990s is not likely to improve the situation.

The Military-Industrial Complex

Reciprocal consent between the L.D.P. and big business has helped to lay the foundations for a Japanese military-industrial complex. This was, oddly enough, the very thing that Occupation Forces had been sent in to destroy. Strong government-industrial links have assisted in its re-emergence. In 1979, half of defense procurements went to five companies: Mitsubishi Heavy Industries, Mitsubishi Electric, Kawasaki Heavy Industries, Ishikawajima Harima Heavy Industries, and the Toshiba Corporation.[45] In order to encourage self-sufficiency in military equipment and arms, the Defense Agency tends to purchase from Japanese companies wherever possible, even if it means paying double the cost of the same thing made overseas. The Japanese

44. Kent E. Calder, "Japan in the Emerging Global Political Economy," in *Power, Economics, and Security: The United States and Japan in Focus,* ed. Henry Bienen (Boulder, Colo.: Westview Press, 1992), 226–53, esp. 256–57.

45. McIntosh, *Japan Re-armed,* 52–53.

government thus subsidizes the Japanese arms industry and this has resulted in arms self-sufficiency: 99 percent of naval ships are Japanese-made; as are 89 percent of aircraft; 87 percent of ammunition; and 83 percent of firearms.[46] The Defense Agency and its main suppliers have come to be known as the 'defense family.' The biggest twenty suppliers take 75 percent of defense orders, and they do not have to make competitive bids on contracts. Being part of a 'family', there are obligations on the companies as well. Three hundred forty retiring self-defense force officers were hired by these companies between 1975 and 1980.[47]

Japan's arms industry, restricted by the 1967 Satô government ban on the export of weapons, is limited to the domestic market, and makes up only one-third of 1 percent of all Japanese production—i.e., about $U.S. 15 billion a year, or the equivalent of Japan's tire industry. If the ban were lifted, it is estimated that Japan could take control of 60 percent of the market for world warship construction, 40 percent of military electronics, 25–30 percent of the aerospace industry, and 46 percent of tanks and motorized artillery. In the meantime, Japan does export arms. In 1984, it became the world's twentieth largest arms exporter. Japanese companies are allowed to export products with military applications only if civilian applications can be found, but a Japan Broadcasting Company (N.H.K.) survey of thirty electronic companies in the United States showed that one in three use Japanese electronics in military products.[48]

On 29 March 1990, the International Peace Research Institute in Stockholm announced the top one hundred arms manufacturers in the world. Despite prohibitions on the export of arms, six Japanese companies were included among them:

23. Kawasaki Heavy Industries (sales of $U.S. 2.23 billion)
28. Mitsubishi Heavy Industries
58. Mitsubishi Electric
63. Toshiba
68. Ishikawajima-Harima Heavy Industries
88. N.E.C.[49]

Despite Solingen's predictions (chap. 1, this volume) that military-industrial complexes are likely to decline, the threat posed by China and North Korea will not disappear. Should Japan decide to lift its ban on exports of military

46. Peter Hartcher, "The Arsenal Ready and Waiting in the Pacific," *Sydney Morning Herald,* 6 April 1988, 17.

47. Hartcher, "Arsenal Ready and Waiting," 17.

48. Hartcher, "Arsenal Ready and Waiting," 17.

49. "Sekai heiki meekaa 100 ketsu: Nihon kara 6 sha" ("World's Top 100 Arms Manufacturers: 6 Companies from Japan Included"), *Asahi shinbun,* 30 March, 1990, 1.

technology and pursue overseas markets, it is expected that in the post–cold war world, the defense industries in Britain, France, and the United States will decline even further. World expenditure on arms dropped by 25 percent in 1992, but military spending actually increased in the Middle East, China, Japan, and other parts of Asia.[50] Morishima has pointed out,

> With her enormous production capacity, once Japan is seriously involved in that business [defense] the balance of the world would change greatly, and a new military problem, Japan vs. the West, would again emerge.[51]

The nature of Japanese R&D would change as well.

Nakasone, Physicists, and the State

Nakasone's anticommunist and militaristic vision for Japan has contributed to this state of affairs. His career reflects a blurring of the divisions between the bureaucracy and the L.D.P., his power constrained by factional politics within his party, and predicated on a commitment to include big business in the policy process.[52] He is noteworthy for having been a new type of Japanese political leader when he was in power during the period 1982–87 for two and a half terms, a prime minister who had a high personal profile, imposing his own stamp of leadership in hitherto weak areas of Japanese policy-making such as foreign affairs. He is also significant in having been one of the few Japanese politicians who has shown a long-term interest in science and technology.

Nakasone was a rising political star from early on, and was the first Japanese politician to show serious interest in atomic energy. He quickly established a reputation for himself as a conservative member of the Diet who was an outspoken nationalist. It is therefore not surprising, given the possible military potential of any developments, that he would pursue nuclear power. His efforts in this regard have enabled Japan to arm itself with nuclear weapons if it ever should wish to do so. This part examines various phases of Nakasone's career as a basis upon which to further explore state-physicist relations.

Interaction with Physicists

As early as 1951, Nakasone visited the physicist Seiji Kaya, making enquiries about nuclear power. He went to the United States around 1953 and spoke

50. Anonymous, "World's Spending on Arms Falls by a Quarter," *The Australian* (16 June 1993): 10.

51. Cited in McIntosh, *Japan Re-armed,* 61.

52. Samuels, *The Business of the Japanese State.*

with Ryôkichi Sagane, at the Lawrence Laboratory at Berkeley, before attending a summer course at M.I.T. on atomic energy. Nakasone apparently stressed to Sagane how important nuclear power was to Japan and how he had hopes of establishing a program.[53] Nakasone's trips overseas were not only fact-finding tours, but ways of extending his network of foreign contacts. His knowledge of Japan's international context was incorporated into the policy-making process. He and other politicians would later find that gaiatsu could be manipulated for domestic change.[54]

In March 1954, after having returned from abroad, Nakasone made a budget request for an additional ¥250 million ($U.S. 694,444) for science and technology, ¥235 million ($U.S. 652,777) of which would be allocated for the construction of a nuclear reactor and ¥15 million towards uranium prospecting.[55] Nakasone was the main promoter of the surprise budget request in the minority political party called Kaishintô (Progressive Party). The party had the casting vote in terms of passing the 1954 budget, and was therefore successful in having the proposal supported by the other two conservative parties, the Liberal Party and the Japan Democratic Party.[56] This was approved in great haste by the House of Representatives, against the wishes of physicists in the Science Council. The Council would have preferred that atomic energy policy be first debated before any allocation was made.[57] Instead, the nuclear budget was government-initiated. This surprise action set the tone for the estranged relationship between government-industry circles and their scientist-critics.

It is not surprising that the bulk of the atomic energy budget was administered by the influential Ministry of International Trade and Industry through a committee which it organized. The government also set up a Preparatory Committee for the Peaceful Use of Atomic Energy, which would lead to the establishment of the Japan Atomic Energy Commission.[58]

Nakasone allied himself with other actors whose interests conformed with his own. He was instrumental in having Matsutarô Shôriki appointed

---

53. Japan Atomic Industrial Forum, *Atomic Energy,* 5; "Dôryoku ro no dônyû: Genden jidai o chûshin to shite" ("The Introduction of a Power Reactor: The Japan Atomic Power Company Days"), roundtable discussion on 10 September 1980, in *Sagane Ryôkichi kinen bunshû (Collection of Writings to Commemorate Ryôkichi Sagane)* ed. Publication Committee (Tokyo: Sagane Ryôkichi kinen bunshû shuppankai, 1981), 253–73, esp. 256.

54. Calder, *Crisis and Compensation,* 121–22.

55. Masanori Ônuma, Yôichirô Fujii, and Kunioki Katô, *Sengo Nihon kagakusha undô shi: Jyô (A History of the Postwar Japanese Scientists' Movement: Part One)* (Tokyo: Aoki Shoten, 1975), 120.

56. Seiji Kaya, "The Activities Shown by Dr. Fujioka at the Initial Stage of the Peaceful Use of Atomic Energy in Japan," *Yoshio Fujioka Commemorative Issue* (Tokyo: Institute for Optical Research, Tokyo University of Education, 1967), 370–72, esp. 371.

57. Yanaga, *Big Business,* 179–80.

58. Kaya, "The Activities Shown by Dr. Fujioka," 371.

minister of state in charge of atomic energy, through his acquaintance with Bukichi Miki, a close adviser to the prime minister. On 26 November 1955, Nakasone and the Joint Diet Atomic Energy Committee made representations to the Cabinet Preparatory Council for the Use of Atomic Energy. It was argued that there was an urgent need for a national policy for atomic energy. Nakasone was supported by three physicists who were present: Seiji Kaya, Yoshio Fujioka, and Sin-itirô Tomonaga.

Nakasone made sure that the J.A.E.C., which was subsequently established, was not an impotent one by having included a section that compelled the prime minister to respect the decisions of the Commission. It seems that such strong wording was unusual in legislation at the time.[59] He made overtures to the physicist Seiji Kaya to encourage Nobel prize winner Yukawa to become a member of the Commission.[60] Big business would be represented by Ichirô Ishikawa, president of the powerful Keidanren (Federation of Economic Organizations), and Hiromi Arisawa would also be a member. While preparing the legislation for atomic energy, Nakasone became convinced of the need for a science and technology agency. One of his colleagues, Diet member Masao Maeda, spearheaded preparations for the establishment of an agency, but Nakasone was able to have considerable input.[61]

Going Up in the World

Nakasone was making his interest in science and technology work to his political advantage. In June, 1959, he became a state minister, director of the Science and Technology Agency for one year, and until July, 1960, chair of the Japan Atomic Energy Commission. During this time he left his mark on science policy. In 1959, he established a private advisory committee of over ten scholars called the *Uchû Kagaku Gijutsu Shinkô Junbi Iinkai* (Committee for the Promotion of Space Science and Technology). One year later, this committee circulated a report entitled *Tômen no uchû kagaku gijutsu kenkyû kaihatsu keikaku* (Immediate Plans for Space R&D) which argued for the establishment of a Space Science and Technology Council within the Prime Minister's Office.[62] In August, 1961, Nakasone became chairman of the L.D.P. Special Committee for Science and Technology. He thus emerged as a

59. Yasuhiro Nakasone, "Kagaku Gijutsuchô setsuritsu made no omoide" ("Memories up until the Establishment of the Science and Technology Agency"), in *Kagaku Gijutsuchô: 30 nen no ayumi (The Science and Technology Agency: A 30 Year History,)* Japanese Government, Science, and Technology Agency (Tokyo: Sôzô, 1986), 94–95. Originally printed in *Kagaku Gijutsuchô: Jûnen shi* (Tokyo: Science and Technology Agency, 1966), 27–28.

60. Japan Atomic Industrial Forum, *Atomic Energy,* 55.

61. Nakasone, "Memories."

62. Tôki Yatô, *Uchû kaihatsu seisaku keisei no kiseki (A History of the Formation of Space Development Policy)* (Tokyo: Kokusai Tsûshin Bunka Kyôkai, 1983), 32.

powerful politician, with a strong focus on science and technology and a distinct liking for international travel!

Japan's Defense

In December 1969, Nakasone was reelected and soon after was made the director general of the Defense Agency, continuing in that position until July, 1971. It was not a particularly popular position for aspiring politicians, no postwar prime minister at that time ever having occupied that position.[63] But Nakasone's long-term concern about Japan's defense meant that the position was one which suited his interests. His interests in science and technology were now at one with his official role of overseeing Japan's defense capabilities.

In July, 1970, U.S. Defense Secretary Laird made a long visit to Japan and Korea. He urged Japan to buy more U.S. military equipment and to exert a stronger military role in East Asia. The question of nuclear weapons was apparently discussed in detail as well.[64] Later that year, Nakasone visited the United States for discussions aimed at negotiating a policy whereby the United States and Canada would enter into a joint partnership with Japan in supplying Japan with know-how for the enrichment of uranium. Nakasone met once more with Laird. There was considerable media speculation throughout the world that the negotiations would herald the emergence of Japan as a military nuclear power. Nakasone left that option open with his noncommital statements that Japan would not do so for as long as the U.S. nuclear deterrent remained a "credible" one.[65]

The Prime Ministership

Shortly before becoming the eleventh L.D.P. president in November, 1982 and thereby prime minister, Nakasone spoke of "Having 'caught up,' we must now expect others to try to catch up with us. We must seek out a new path for ourselves and open it up ourselves. . . ."[66] Science and technology would lead Japan into the twenty-first century.

Nakasone learned from his experience with decision making in science

63. Ken'ichi Nakamura, "Militarization of Post-war Japan," in *Asia: Militarization and Regional Conflict,* ed. Yoshikazu Sakamoto (Tokyo: United Nations University/London: Zed Books, 1988), 81–100, esp. 99.

64. Halliday and McCormack, *Japanese Imperialism Today,* 94–96.

65. Halliday and McCormack, 94–96.

66. The manuscript was published as Nakasone, "Toward a Nation of Dynamic Culture and Welfare," abridged trans. of Yasuhiro Nakasone, "Takumashii bunka to fukushi no kuni o," *Seiron* (January 1983): 26–37.

and technology policy, and sought to appoint a large number of councils and committees to deliberate upon often predetermined policy recommendations. Membership tended to include academics, businessmen, and other opinion-makers who generally agreed with Nakasone's views on matters such as education, defense, and the economy. This technique enabled Nakasone to influence policy and to give his ideas an authority courtesy of the eminent members of his various committees.[67]

Nakasone did not forget about science and technology. In a speech to the Diet on 6 February 1984, he spoke of "the achievement of a sophisticated information society [as] an important strategic element in medium and long-term economic development for the twenty-first century."[68] He and his supporters recognized that Japan's position as a major world power in the next century largely depended on nurturing scientific and technical skills, but as Kenneth Pyle has put it, they were aware that "scientific leadership will require political leadership as well."[69] While some physicists were aware of this, many were not in the position of achieving this within the confines of the academic community. The media offered exposure for their opinions, as did various councils and committees, but as Nakasone knew all too well, the latter could be manipulated to suit the interests of politicians and bureaucrats.

Nakasone remained in office for five years. His prime ministership encouraged projects such as the I.N.S. in telecommunications, the fifth-generation computer project, and the ambitious M.I.T.I. technopolis plan.[70] National strength in Nakasone's eyes also required a stronger military capability, but as his faction was only the fourth largest,[71] he had difficulty in realizing policies for national rearmament. Nevertheless, the military budget rose by 5.2 percent each year, when overall government expenditures went up by only 1.7 percent. In July, 1986, after what was arguably the L.D.P.'s greatest postwar election victory, Nakasone was able to gain a one-year extension to his term of office as president of the Liberal Democratic Party and as prime minister. Given this mandate to govern, the self-imposed 1 percent of G.N.P. barrier on defense spending was finally exceeded in 1987.[72]

Related to the threat posed by an emerging military-industrial complex is the fact that Nakasone has helped to fashion a policy-making structure that ties science and technology firmly to national interest. In 1952, advisory councils established under the aegis of a ministry numbered 165. By 1969, the figure

---

67. Kenneth B. Pyle, "In Pursuit of a Grand Design: Nakasone betwixt the Past and the Future," *The Journal of Japanese Studies* 13, no. 2 (Summer 1987): 243–70, esp. 253.

68. Pyle, "In Pursuit of a Grand Design," 255.

69. Ibid., 254.

70. Ibid., 255.

71. Nakamura, "Militarization of Post-war Japan," 82.

72. Calder, *Crisis and Compensation,* 426.

had increased to 243. As the agenda for such advisory bodies is determined by the particular ministry in question, topics tend to be narrowly defined and are apt to lead to recommendations that seem predetermined.

It is clear that Nakasone had a major impact on science policy. He reminded physicists that they served only as advisers to the government. Politicians, bureaucrats, and business leaders would ultimately determine what strategies the nation adopted. His longstanding interest in science policy reflects more his nationalistic character and political priorities, and to a lesser extent an interest in economic returns.

State-Scientist Relations

This chapter has revealed patterns of political relations involving the scientific community in Japan. The first part showed how both the internal structures of the political economic system and the global context (security concerns and trade) influence interaction between the state and scientists. We have examined some of the domestic political institutions and policy-making processes. The Allied Occupation, growing U.S.-Japan relationship, onslaught of the cold war, and 'catching-up' with Western technology all were conceptions of the political economy that impacted upon Japan's scientific and technological capabilities and established gaiatsu as a major factor in policy-making. Nakasone's career, discussed in the second section, enabled us to access Japan's experience in terms of real actors and individual actions. The former prime minister was a major player whose very actions reflect the political economy within which he was immersed.

The Domestic Structure of Scientific Activity

What impact have physicists had in the negotiation of science policy? The postwar activities of Japanese physicists can be interpreted in many ways. We have seen how physicists emerged as an influential elite immediately after World War II, and how physicists have helped to shape the state policy-making structures. They expressed their concerns over science policy, the development of civilian nuclear power, and the related issues of arms control and economic development—irrevocably politicizing the scientific community. Nakasone's career reminded us that the state's involvement in international political and economic systems—through the U.S.-Japan relationship and with the market via reciprocal consent—was reflected in the development of Japan's science and technology.

T. Dixon Long has tended to dismiss the impact of physicists on Japan's postwar development. He has judged the Science Council of Japan as being not particularly effective, but also sees the more government-aligned Science and Technology Council as lacking in expertise and without sufficient support from the scientific community. He is critical of scientists for the way they have clashed with the Japanese government and he tends to view their protests over such issues as radiation from American nuclear submarines at U.S. bases in Japan in a poor light. Similarly he considers their desire to build a large-scale particle accelerator (the type of facility that U.S. and European physicists enjoyed access to) as misguided.[73]

Rather than merely dismiss such desires, it is more useful to come to an understanding as to what motivates them. National prestige and the attraction of being able to participate at the forefront of research are major factors in the pursuit of physics.[74] When it comes to projects such as the giant superconducting supercollider in Texas, factors such as gaiatsu resulting from trade frictions are just as important as individual ambitions in encouraging Japan to invest huge sums in the project. It is in massive projects such as this that the interests of scientists and the state converge.

Although many physicists associated with the Science Council of Japan have had an adversarial relationship with the state, there have been others whose roles have been more technocratic in nature. Long has greatest respect for the scientists Kankuro Kaneshige and Seiji Kaya, who tended to have views that accommodated those of the government and business. Physicists with more liberal views such as Shôichi Sakata and Sin-itirô Tomonaga, who both showed great concern for issues such as freedom of research, are deemed to have had little influence on national policy-making. This is despite Tomonaga having served two terms as president of the Science Council, and wielding influence as a result of the award of the Nobel Prize for physics in 1965. While it is clear that physicists have tended toward the pursuit of curiosity-driven research and resisted pressure from government and business to adopt strategies aimed at securing practical benefit from their creativity as rapidly as possible, Long's narrow evaluation of the impact of physicists neglects the importance of the research infrastructure which they helped to establish; the role of other physicists, such as Satio Hayakawa, on various councils; and most worrisomely, adopts the idea of industrialization at all costs. Many physicists, with a keen sense of responsibility, acted as social spokespersons. Some may see their greatest contribution to science policy and Japan's postwar modernization as in how, according to Long, "the Science Council has embarrassed and thwarted the government in several ways."[75]

---

73. T. Dixon Long, "The Dynamics of Japanese Science Policy," in *Science Policies of Industrial Nations,* ed. T. Dixon Long and Christopher Wright (New York: Praeger Publishers, 1975), 133–68, esp. 155–56.

74. This is discussed in Shigeru Nakayama, *Science, Technology and Society in Postwar Japan* (London: Kegan Paul International, 1991), 76–78.

75. Long, 156–57.

Long's interpretation needs to be modified if we view Japan's development in the wider sense of the needs and social welfare of the people. Their contributions in terms of Japan's pacifist stance, growing research infrastructure and international outlook in science may take time to be fully appreciated. The physicists examined sat on important Science Council and government committees, established major research institutes and laboratories, helped determine the manner in which they were run, contributed to the public image of science, and determined the direction of basic science through their deliberations.

The very organization of science and technology in Japan is symptomatic of how big business, the bureaucracy, and politicians have tried to determine that science can be something akin to what it was last century: in service to the state. There is evidence that some state structures have conformed to private interests.[76] To the dismay of physicists, by the time Nakasone became prime minister in the 1980s, a multilayer structure for policy-making and the management of science had been reinstalled, a system just as complex as that which had existed during World War II when there were attempts to integrate science with big business and the state. Science and technology were remobilized for the nation, and there has been a growing militaristic aspect to their growth. Crucial parts of the system were put into place in response to the demands of physicists for democratic representation (e.g., the Science Council of Japan), and also in order to thwart them by using a separate structure more attuned to the needs of the L.D.P., bureaucracy, and big business (e.g., the Council for Science and Technology). Such bodies and the ministries to which they answer, now form a complex web of organizations involved in the decision-making process. While there is physicist input into this web, it can be shown that in many cases this system served to duplicate or exclude physicist-led committees, councils, and research institutes. This wasteful duplication, the result of power struggles and factionalism, has made for a pluralistic structure in various fields of science. Often the divisions are along the lines of basic versus applied and academic versus commercial, reflecting fundamental differences of opinion as to what science constitutes. State power and private power have often enhanced each other. The pluralistic structures of science suggest that this is how a balance is struck between state jurisdiction and private control.

This is not surprising given that, unlike in the United States, the state is not the main patron of basic research. Nevertheless, the state has been able, through close ties between big business, the bureaucracy, and the L.D.P., to exercise considerable control over private investment of science and scientific relations with the outside world. While physicists have enjoyed a high degree

76. See Samuels, *The Business of the Japanese State,* 256.

of scientific freedom at universities, they attempted to extend this to include management of inter-university research institutes and large experimental facilities, often not meeting with success. Yet their usefulness as policy advisers/legitimators has meant that insiders are still able to contribute to the policy-making process and outsiders/dissenters are relegated to forums such as the Science Council or the mass media.

The relatively weak links between university-based physicists and private industry have been compensated for by the high degree of privately funded research and strong government-industry links, especially in the case of M.I.T.I. Left-wing physicists such as Sakata and Taketani tended to be theoreticians and had no great need for research funding. They did, however, feel compelled to act as public spokespersons. In pursuit of their vision for postwar Japan, the physicists sought to implement 'objective', democratic policy-making which would somehow be decided upon by 'disinterested' parties for the public good. But such aspirations threatened the authority of the state. The resulting structure has tended to lend legitimacy for often predetermined decisions. While it suggests science policy remains in service to public interest, the reality is that science policy is tied to the agenda of the power elite and the management and administration of science is structured to favor their interests. Japan's emphasis on economic growth and its need to 'catch up' dictated that many of the concerns expressed by scientists would have to take second place to commercial considerations and questions of national security.

Experimental physicists, such as Sagane, were sought out by the state and industry to assist in building the infrastructure for the development of civilian nuclear power. They found that public policy and management of innovation have emphasized the utilization of foreign technology rather than the creation of internal technology. Given that technological borrowing will be more difficult in the future, and increasing competition from newly industrialized economies threatens Japan's export success, Japanese firms will need to respond to the challenges posed by the need for innovation. One way may be for the Japanese to make better use of universities (the source of engineers and scientists) as research institutions; more military research may be an option, as might be a more concerted push for basic research in government and corporate laboratories.[77]

The International Context

The Allied Occupation redefined the Japanese political system, linking scientific activities strongly to the global system. While military-related scientific

77. Mowery and Rosenberg, *Technology and the Pursuit of Economic Growth,* 235–37.

research is very low when compared with the United States, there is the potential for rapid growth. Nakasone showed how security considerations encouraged him to pursue atomic energy. The atom provided a possible route to energy self-sufficiency and a future means of rearmament.

The pluralistic nature of the organization of R&D and the complex policy-making structure reflect the state's participation in the international economic system. In the case of Japan, there has clearly been an emphasis on maintaining a competitive industrial basis over a military one. The recent internationalization of Japan's domestic market and the decline of state control over capital and technology have made for weaker government control overall and greater potential for gaiatsu to influence policy. But a word of caution. Conflict between ministries of the type discussed in this chapter have affected the ability of bureaucrats to enforce policies. Even with U.S. pressure on the Japanese government to improve access to Japan by foreign investment, the ability of public policy to facilitate this by changing industrial organization becomes increasingly difficult.[78] It is thus clear that domestic political and economic structures, and their international context, serve to shape scientist-state relations in a complex manner, but are there general patterns of interaction?

State-Scientist Relations: Which Type?

Japanese intellectuals have been characterized as being able to influence government as insiders, consultants or independents, and among intellectuals, physicists have been prominent. Japanese physicists had great moral authority, backed by science whose authority was unquestionable.[79] Nishina was an insider who had a considerable impact on decision-making processes. He was in contact with the military during the war, and in close contact with S.C.A.P. after it. Nishina urged Japanese physicists to translate their authority into political clout and to use the scientific method to solve Japan's problems. Both Nishina and Sagane felt that science could be used for practical ends such as medical research and nuclear power. For them, physics had no true path of its own, rather, it lent itself to increasing opportunities for integration into industry.

The experts Yukawa and Tomonaga were more consultants than insiders, providing appropriate advice, information, and credibility when needed. Both felt that science could contribute to society by adding to the bank of knowledge. To their adoring public, the views of Nobel Prize winning physicists

78. Taylor and Yamamura, 31, 37.

79. Herbert Passin, "Intellectuals in the Decision Making Process," in *Modern Japanese Organization and Decision-making,* ed. Ezra F. Vogel (Berkeley: University of California Press, 1975), 251–83.

seemed best. Even the political left coveted their authority, especially when displayed on public statements or petitions calling for peace. The relatively apolitical Yukawa and Tomonaga did not need to or care to take to the streets, for the idea of peace provided a rallying point to organize Japanese physicists and to bring together other Nobel Prize winners.

Sakata and Taketani viewed the role of science in terms of Marxism being able to change society. They sought to influence policy-making through the normal channels available in a democracy: through their writings, the mass media, and input into scientists' movements. As social spokespersons, they shaped public opinion, contributed to the activities of social movements, and helped forge the attitudes of scientists and intellectuals. While Sakata was respected for his physics, his anti-establishment character excluded him from much in the way of official government responsibilities.

Compared to Sakata and Taketani, Hayakawa and Sagane were relatively neutral physicists, insiders who cooperated with the establishment to realize their very many projects. Spanning cyclotrons to reactors, Sagane's entire career was concerned with U.S.-Japan technology transfer. Upon his return from many years at Berkeley, he started a new career as a *goyôgakusha,* a scholar who is paid to provide services to the government.[80] Unlike his other physicist-colleagues, Sagane differed by being in the direct employ of public corporations, and for that reason thought of in a far less kinder light than those based in universities. Despite a lack of popularity, he served a pivotal role as bridge between government and the estranged academic community. He applied his skills to tasks requested by the government, and influenced the execution of policy rather than determined it. He worked for so-called public corporations, such as J.A.E.R.I. and J.A.P.CO. (Japan Atomic Power Company), reflecting how in the late 1950s and 1960s, the term *public* came to be associated with public authority rather than public discussion *à la* Sakata and Taketani.

In an area such as atomic energy, where powerful pressure groups were involved, experts would be subordinate to the wishes of politicians such as Nakasone. The second part of this chapter outlined the impact of Nakasone and the disparity between his approach to science and that of many of the physicists mentioned. This gap can be attributed to both political and conceptual differences as to what science should be. Nakasone was part of the establishment and arguably the single most important political figure in the postwar period to have dealt extensively with science and technology. His administration is largely credited with the push toward the rearmament of Japan and the development of the concept of Japan as an information society.[81]

---

80. Passin, "Intellectuals in the Decision Making Process," 265.

81. Pyle, "In Pursuit of a Grand Design," 255.

Anti-establishment intellectuals often have no choice but to try to influence policy making as independents. It is obvious that insiders and consultants tend to come from those who do not oppose the establishment and who are politically neutral, but their potential to influence government policy is considerable, especially when there is a very small pool of people with the requisite skills. As insiders, they can supply information and formulate policy. Especially for technical matters with little political ramifications, insiders such as Sagane and, to a lesser extent, Hayakawa were particularly useful. Hayakawa's role in establishing various research institutes and laboratories was a means by which Japan could become a member of the international community at a time when scientific dialogue was sadly lacking. He actively sought to carry on the leadership roles carved out by Nishina and Tomonaga. His experience provides further evidence of how big science is dependent on bureaucrats in the Ministries of Education and Finance, and Nakasone's career indicates that bureaucrat/ex-bureaucrat domination of the policy-making process is no accident.

The above discussion suggests that there is a happy convergence of interests—to use Solingen's term in the introductory chapter—between the state and physicists who are insiders (especially those working in the private sector and in nonuniversity, government research institutions). At the same time, however, there has been a strong, underlying tension between the state and left-wing scientists, resulting in what can be aptly described as ritual confrontation along political lines. The state has sought to privatize this conflict via the policy-making structure, rendering dissident voices increasingly passive. Both models cf state-scientist interaction coexist in Japan, their degree of applicability dependent upon the subgroup under examination.

BIBLIOGRAPHY

Akhavan-Majid, Roya. "Telecommunications Policy-making in Japan, 1970–1987: A Case Study in Japanese Policy-making Structures and Process." Ph.D. diss., University of Minnesota, 1988.

Benjamin, Roger, and Ori, Kan. *Tradition and Change in Postindustrial Japan: The Role of the Political Parties*. New York: Praeger, 1981.

Bienen, Henry, ed. *Power, Economics, and Security: The United States and Japan in Focus*. Boulder, Colo.: Westview Press, 1992, 226–53.

Calder, Kent E. *Crisis and Compensation: Public Policy and Political Stability in Japan: 1949–1986*. Princeton: Princeton University Press, 1988.

———. "Japan in the Emerging Global Political Economy." In *Power, Economics, and Security*. See Bienen 1992, 226–53.

Fukui, Haruhiro. "Studies in Policymaking: A Review of the Literature." In *Policymaking in Contemporary Japan*. See Pempel 1977, 22–59.

Halliday, Jon, and McCormack, Gavan. *Japanese Imperialism Today: 'Co-Prosperity in Greater East Asia'*. New York: Monthly Review Press, 1973.
Harries, Meirion, and Harries, Susie. *Sheathing the Sword: The Demilitarisation of Japan*. London: Hamish Hamilton, 1987.
Hartcher, Peter. "The Arsenal Ready and Waiting in the Pacific." *Sydney Morning Herald* 6 April 1988, 17.
Heiduk, Gunter, and Yamamura, Kozo, eds. *Technological Competition and Interdependence: The Search for Policy in the United States, West Germany, and Japan*. Seattle, Wash.: University of Washington Press/Tokyo: University of Tokyo Press, 1990.
Hirosige, Tetu. *Sengo Nihon no kagaku undô (Postwar Japan Science Movement)*. Tokyo: Chûô Kôronsha, 1960.
Hogwood, Brian W., and Gunn, Lewis A. *Policy Analysis for the Real World*. Oxford: Oxford University Press, 1984.
Hunter, Janet E., comp. *Concise Dictionary of Modern Japanese History*. Berkeley: University of California Press, 1984.
Institute for Optical Research. *The Yoshio Fujioka Commemorative Issue*. Tokyo: Tokyo University of Education, 1967.
Itoh, Hiroshi, ed. and trans. *Japanese Politics—An Inside View: Readings from Japan*. Ithaca, N.Y.: Cornell University Press, 1973.
Japan Atomic Industrial Forum. *Nihon no genshiryoku: 15 nen no ayumi, jyô (Atomic Energy in Japan: A 15 Year History, Part 1)*. Tokyo: Japan Atomic Industrial Forum, 1971.
Japanese Government. *Kagaku gijutsu hakusho (Heisei yonnen ban) (Science and Technology White Paper)*. Tokyo: Ministry of Finance Printing Office, 1992.
Japanese Government, Science and Technology Agency, *Kagaku Gijutsuchô: 30 nen no ayumi (The Science and Technology Agency: A 30 Year History)*. Tokyo: Sôzô, 1986.
Johnson, Chalmers. *M.I.T.I. and the Japanese Miracle: The Growth of Industrial Policy, 1925–1975*. Stanford: Stanford University Press, 1982.
Kaya, Seiji, "The Activities Shown by Dr. Fujioka at the Initial Stage of the Peaceful Use of Atomic Energy in Japan." In *The Yoshio Fujioka Commemorative Issue*. See The Institute for Optical Research 1967, 370–72.
Koh, Byung Chul. *Japan's Administrative Elite*. Berkeley: University of California Press, 1989.
Long, T. Dixon. "The Dynamics of Japanese Science Policy," In *Science Policies of Industrial Nations*. ed. T. Dixon Long and Christopher Wright. New York: Praeger Publishers, 1975, 133–68.
McIntosh, Malcolm. *Japan Re-armed*. London: Frances Pinter, 1986.
Makin, John H., and Hellmann, Donald C., eds. *Sharing World Leadership?: A New Era for America and Japan*. Washington, D.C.: American Enterprise Institute for Public Policy Research, 1989.
Misawa, Shigeo. "An Outline of the Policy-making Process in Japan." In *Japanese Politics—An Inside View*. See Itoh 1973, 12–48.
Mowery, David C., and Rosenberg, Nathan. *Technology and the Pursuit of Economic Growth*. Cambridge: Cambridge University Press, 1989.

Murakami, Hyôe. "The Making of a Prime Minister." In *Politics and Economics in Contemporary Japan.* See Murakami and Hirschmeier 1983, 3–20.

Murakami, Hyôe, and Hirschmeier, Johannes, eds. *Politics and Economics in Contemporary Japan.* Tokyo: Kodansha International, 1983.

Muramatsu, Michio, and Krauss, Ellis S. "Bureaucrats and Politicians in Policymaking: The Case of Japan." *American Political Science Review* 78 (1984): 126–46.

Nakamura, Ken'ichi. "Militarization of Post-war Japan." In *Asia: Militarization and Regional Conflict.* See Sakamoto 1988, 81–100.

Nakasone, Yasuhiro. "Kagaku Gijutsuchô setsuritsu made no omoide" ("Memories up until the Establishment of the Science and Technology Agency"). In *The Science and Technology Agency.* See Japanese Government, Science and Technology Agency 1986, 94–95.

Nakayama, Shigeru. *Science, Technology and Society in Postwar Japan.* London: Kegan Paul International, 1991.

Ônuma, Masanori, Fujii, Yôichirô, and Katô, Kunioki. *Sengo Nihon kagakusha undô shi: Jyô (A History of the Postwar Japanese Scientists' Movement: Part One).* Tokyo: Aoki Shoten, 1975.

Passin, Herbert. "Intellectuals in the Decision Making Process." In *Modern Japanese Organization and Decision-making.* See Vogel 1975, 251–83.

Peck, Merton J., with Tamura, Shûji. "Technology." In *Asia's New Giant: How the Japanese Economy Works,* ed. Hugh Patrick and Henry Rosovsky. Washington, D.C.: Brookings Institution, 1976: 525–85.

Pempel, T. J. "The Bureaucratization of Policymaking in Postwar Japan." *Midwest Journal of Political Science:* 18 (1974): 647–64.

———, ed. *Policymaking in Contemporary Japan.* Ithaca: Cornell University Press, 1977.

———. "The Unbundling of 'Japan, Inc.': The Changing Dynamics of Japanese Policy Formation." *Journal of Japanese Studies* 13, no. 2 (Summer 1987): 271–306.

Publication Committee, ed. *Sagane Ryôkichi kinen bunshû (Collection of Writings to Commemorate Ryôkichi Sagane).* Tokyo: Sagane Ryôkichi kinen bunshû shuppankai, 1981.

Pyle, Kenneth B. "In Pursuit of a Grand Design: Nakasone betwixt the Past and the Future." *Journal of Japanese Studies* 13, no. 2 (Summer 1987): 243–70.

———. "The Burden of Japanese History and the Politics of Burden Sharing." In *Sharing World Leadership?* See Makin and Hellmann 1989, 41–77.

Ronayne, J. *Science in Government.* Melbourne: Edward Arnold, 1984.

Sakamoto, Yoshikazu, ed. *Asia: Militarization and Regional Conflict.* Tokyo: United Nations University/London: Zed Books, 1988.

Samuels, Richard J. *The Business of the Japanese State: Energy Markets in Comparative and Historical Perspective.* Ithaca, N.Y.: Cornell University Press, 1987.

Schaller, Michael. *The American Occupation of Japan: The Origins of the Cold War in Asia.* New York: Oxford University Press, 1985.

———. *Douglas MacArthur: The Far Eastern General.* New York: Oxford University Press, 1989, 1990.

"Sekai heiki meekaa 100 ketsu: Nihon kara 6 sha" ("World's Top 100 Arms Manufacturers: 6 Companies from Japan Included." *Asahi shinbun* 30 March 1990, 1.

Tamamoto, Masaru. "Japan's Search for a World Role". In *Power, Economics, and Security.* See Bienen 1992, 226–53.

Taylor, Sully, and Yamamura, Kozo. "Japan's Technological Capabilities and Its Future: Overview and Assessments." In *Technological Competition and Interdependence.* See Heiduk and Yamamura 1990, 25–63.

Thayer, Nathaniel B. *How the Conservatives Rule Japan.* Princeton: Princeton University Press, 1969, 1973.

Titus, David Anson. *Palace and Politics in Prewar Japan.* New York: Columbia University Press, 1974.

Vogel, Ezra F., ed. *Modern Japanese Organization and Decision-making.* Berkeley: University of California Press, 1975.

Welfield, John. *An Empire in Eclipse: Japan in the Postwar American Alliance System: A Study in the Interaction of Domestic Politics and Foreign Policy.* London: Athlone Press, 1988.

Westfall, Catherine Lee. "The First 'Truly National Laboratory': The Birth of Fermilab". Ph.D. diss., Michigan State University, 1988.

Wilkie, Tom. *British Science and Politics since 1945.* Oxford: Basil Blackwell, 1991.

"World's Spending on Arms Falls by a Quarter." *The Australian* 16 June 1993, 10.

Yanaga, Chitoshi. *Big Business in Japanese Politics.* New Haven: Yale University Press, 1968.

Yatô, Tôki. *Uchû kaihatsu seisaku keisei no kiseki (A History of the Formation of Space Development Policy).* Tokyo: Kokusai Tsûshin Bunka Kyôkai, 1983.

Yuasa, Mitutomo. *Kagakushi (The History of Science).* Tokyo: Tôyô Keizai Shinpôsha, 1961.

People's Republic of China: Between Autarky and Interdependence

Wendy Frieman

Introduction and Background

The search for generalizations about scientific behavior across countries is irresistible given the role that science and technology plays in the modern world. The ability to predict or understand state-science relations as a function of a nation's form of government, economy, or foreign relations would be a significant step forward in comparative analysis. However, most studies of Chinese science treat the subject in its own context—China is assumed to be a special case, inherently incomparable to other societies.[1] Why is it so difficult to predict state-science relations in China based on norms in other countries? Why is the political economy of Chinese science, as a whole, not a strong base for generalizations about other developing countries?

Several obstacles confront the use of a China case in comparative analysis of scientists and the state. The first is raw data on state-science interactions. Much of what is reported on how the Chinese currently formulate a national science agenda, allocate research funds, reward scientific endeavor, or weigh tradeoffs between basic and applied research, is anecdotal, unsystematic, and virtually impossible to verify. This is also a problem for trying to define and characterize state-science relations in the 1950s, 1960s, or 1970s. Evidence exists, largely in the form of policy pronouncements and debates, to suggest that Chinese leaders have preferred or promoted different models of scientific organizations at different times, but the data to support how science was actually administered during those periods is sparse.[2] As a result, just

1. See Denis Fred Simon and Merle Goldman, eds., *Science and Technology in Post-Mao China* (Cambridge, MA: Harvard University Press, 1989); Richard P. Suttmeier, *Research and Revolution* (New York: Lexington, 1974; reprint ed., Institution Press, Stanford: Hoover, 1980); Leo Orleans, ed., *Science in Contemporary China* (Stanford: Stanford University Press, 1980); *Science and Technology in the PRC* (Paris: OECD, 1977); Joint Economic Committee, Washington DC: Government Printing Office, 1992 *China's Economic Dilemmas in the 1990's,* 1: 527–644.

2. See n. 1.

characterizing state-science relations in China, much less understanding their causes and effects, is an ambitious task.

An equally serious problem is that within China there are many different scientific communities and, correspondingly, different models for state-science interaction. China has over 10,000 science and technology (S&T) research institutes.[3] Estimates of the size of the science and technology workforce (professionals with a Bachelor of Science degree) in China range from 300,000 to 400,000.[4] Chinese definitions attached to these numbers are imprecise, and therefore it is impossible to know if these figures include university personnel, or only the industrial S&T workforce. In any event, the S&T workforce is a diverse and heterogeneous one. These individuals work in vastly different environments with different cultures and even value systems. This system is uneven in quality, even within single bureaucratic channels. China has a large number of "low end" research organizations, a tier of institutes that are medium-range and struggling to move up, and a number of isolated islands of excellence. Thus it makes little sense to draw conclusions about the process of science without paying some attention to the outcome.

Finally, Chinese state-science relations have shifted over time. One reason is the degree to which the structure of China's economy and society have been changing for the past fifteen years. When the structure of a nation's political and economic activities and its role in the international environment are in a state of flux, it stands to reason that state-science relations will also be in transition. During the early 1980s the Chinese appeared to be moving closer to a model of state-science interactions common in many of the industrialized countries.[5] However they now appear to be caught between the need to reap the benefits they believe to be associated with the noncompetitive centrally planned model, and the inability to endure the associated risks.

Nevertheless, China is a critical case study in a comparative volume, for many of the same reasons that it is a difficult one: China's S&T resources and potential are vast. The Chinese have identified modernization of science and technology as a national objective. China has a long tradition of scientific excellence and respect for intellectual pursuit. And, finally, several leading scientific figures have begun to play a political role over the past several years.

This chapter is divided into two sections. The first is an attempt to define the political economy of Chinese science by addressing the individual compo-

---

3. Li Xinnan, "S&T System Reform Studied," *Zhongguo Keji Luntan*, January 1, 1991, 36–38, translated in *Joint Publication Research Service: China Science and Technology*, CST-91-012, June 4, 1991, 8–11.

4. Jin Zhouying, "High-Tech Enterprise Development in the Year 2000 Discussed," *Keji Ribao*, October 22, 1990, 1, translated in *JPRS Report: China Science and Technology*, CST-90-030, 58–61.

5. See Solingen in this volume.

nents that make up the system. The past ten years have provided Western scholars with the opportunity to get a glimpse of how the system actually functions at the working level. These realities have often been inconsistent with (or at least different from) the depiction of Chinese science that appears in the press and forms the basis for popular conceptions about how science is conducted in China. The involvement of scientists in political and nonscientific issues is treated separately because it warrants special consideration. A short concluding section summarizes the individual components of the political economy of Chinese science and offers some thoughts on the degree to which the Chinese case corresponds to international patterns.

If it were possible to generalize about the effects of different forms of political and economic organization on state-science relations, the China case should be fairly straightforward. China could be expected to conform to patterns of state-science relations in other large, centrally planned economies with centrally controlled political systems in which the following characteristics are dominant:

Science is primarily a public good that serves the political function of reinforcing the context in which it operates.

Scientific theory is allowed to function only to the extent that it grants legitimacy to (or does not challenge) that context.

The use of authority is the primary form of control; exchange and persuasion are relatively limited.

Most research is conducted in state-run institutes (rather than in universities or industry).

There is generous support for science.

There is a built-in resistance to basic research.

State science relations are characterized by ritual confrontations and deadly encounters more than happy convergence or passive resistance.

In periods of noninvolvement in international security conflicts, military R&D is deemphasized, the need for secrecy is down-played, and there is more receptivity to international interdependence.

In periods of low integration with the international economic community, the state prefers commercial over military research, and lower levels of scientific interdependence.

There is less receptivity to international and transnational actors than in a market economy.[6]

It will be shown that only a few of these generalizations hold true for China over time. The realities of the research environment, from the use of rewards

6. Ibid.

to the involvement of scientists in political affairs, reflect the degree to which the Chinese experience differs from other socialist countries, centrally planned economies, and developing nations.

The Political Economy of Chinese Science

In this section the key components of the political economy of science in China are addressed one by one. Despite the increased access of the past ten years, characterizing this environment is still a task that relies to some measure on conjecture and speculation.

The Organization and Funding of Chinese Science

How research is organized and funded is a critical component of the political economy of science, as it affects form, content, and output of scientific endeavor. It is integral to the question of how the state controls the scientific community, since the formal organizations of scientific activity constitute the mechanisms of control. Since the Chinese system is in the process of change, some reference is made here to earlier periods.

Chinese scientists work in one of four "spheres" or research communities: the Chinese Academy of Sciences (CAS), the universities, the research institutes subordinate to the various industrial ministries, or the military (see fig. 1). Because China is essentially a centrally planned socialist economy, each of these spheres of activity is owned and controlled by the state. Nevertheless, each has a different approach, agenda, organization, almost an entirely different internal political economy. The high degree of differentiation within the system only underscores the difficulty of characterizing Chinese state-scientist relations overall as "happy convergence," "passive resistance," "ritual confrontation," or "deadly encounter," as defined by Solingen. During a given period, depending on which sector of Chinese science is of interest, state-scientist relations could probably be characterized as all of the above. The different research communities also vary substantially in the content of their research foci: the universities are more oriented toward basic research; the industrial ministries are more oriented toward applied research; and the Academy is able to maintain, overall, a balance between the two. It is worth noting, furthermore, that China's R&D system is comprehensive in scope even if it is uneven in quality. This perhaps constitutes a significant difference from other developing countries who, constrained by financial and human resources, do not attempt broad coverage.

The CAS is comprised of 125 research institutes that have traditionally housed China's top scientific minds. The organization was founded in the 1920s (when it was called the Academica Sinica) and in many respects resembles its Russian counterpart. This is the Chinese scientific elite. Scientists in

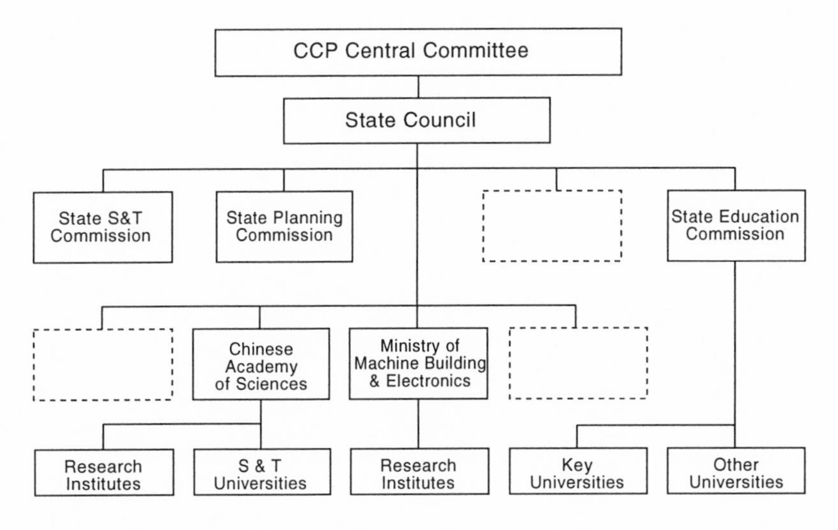

Note: Several other ministries, commissions, and institutes exist that are not depicted here, as indicated by the empty boxes.

Fig. 1. China's S&T organization

these institutes have traditionally had access to key resources: the best graduates, the most sophisticated laboratory equipment, the most comprehensive libraries, and so forth.

The university sector is much larger, with a broader range of quality. There are about 1,060 universities in China, of which 100 are dedicated to science and technology. There is a total of 385,000 faculty across all disciplines, of which 161,000 are either scientists or engineers.[7] Overall, university scientists have been less fortunate in terms of both finance and prestige. University science departments, with very few exceptions (such as Qingdao and Fudan), have had to operate on tight budgets. In addition, the educational establishment was the target of political struggle for a ten-year period, during which time many university scientists were publicly criticized, removed from their positions, and "sent down" to do manual work in the countryside.

The third community is composed of research institutes under the jurisdiction of the industrial ministries, of which there are over 800, and research institutes at the county level, of which there are about 7,500.[8] These are huge organizations about which very little is known. They vary enormously in quality and sophistication. Many of these institutes resemble factories more than research organizations.

7. Zhu Liming, "Colleges, Universities Well Suited to Develop High-Tech Industry," *Kueyuan Guanli,* no. 2, March 1990, 55–57, translated in *JPRS: China Science and Technology,* CST-90-030, 58–61.

8. See n. 4.

One item of interest in a comparative study is how scientists choose their research agenda or, put another way, how the state ensures that national scientific and technological goals are met. From 1949 until 1985 most science establishments in China received block funding, a set budget, from the central government in Beijing, along with a set research agenda. The major decisions were made by the central government, and scientists were either coopted or coerced into adherence to that agenda. Ideological incentives were predominant, although there were sporadic attempts (in the early 1950s and again in the early 1960s) to improve living conditions for scientists and to use material, in addition to normative, rewards. The administration of science resembled the Soviet system on which it was modeled: scientists participated in the discussion about that agenda, but the party always played a dominant role, with the result that many decisions were made for other than scientific reasons, especially during periods of political upheaval and ideological campaigns.

This system changed dramatically in 1985. At that time, five years into the modernization program that was launched following the death of Mao Zedong, the leadership reached a consensus that science and technology had been a major bottleneck to economic modernization. Beginning with the National Science Conference of March 1978, the Deng leadership had assured scientists that they would be free from political persecution, and not constrained by cumbersome political demands. The leadership made numerous efforts to harness the talents and energies of this community for the Four Modernizations program set forth in the early 1980s, an ambitious agenda for progress in agriculture, industry, science and technology, and national defense. Living conditions were improved, salaries were raised, and there were serious attempts by the leadership to reintroduce professional standards for promotion and to minimize the political criteria that had dominated scientific research for so long. Nevertheless, performance remained lackluster. At the local level the Party had retained considerable influence, and science and technology personnel were not in positions of real authority. Worse yet, scientific work remained isolated from meaningful application. In the face of this situation, the leadership agreed on several major reforms in the organization and conduct of science and technology of which perhaps the most important was the change in resource allocation. Institutes would no longer be given block grants to cover all their expenses. They would receive enough money only to cover rent and salaries; all other funding would have to be secured through competition for research contracts from factories or grants from central science organizations. These contracts or grants would be made either on the basis of economic viability (in the case of factories) or on the basis of peer review. Subjective, political factors would be eliminated from funding decisions. If they were successful in generating their own research *funds,* institute directors would have more freedom to determine their own research *agendas*. This was an unavoidable consequence of rationalizing the funding process.

The science reforms, in short, paralleled the reforms in industry and agriculture that had begun to introduce market forces into China's socialist economy. The science reforms had the effect of putting scientists and technical experts in charge, de facto if not de jure of critical resource allocation decisions that would affect China's future. If fully implemented, these reforms would substantially alter the political economy of Chinese science and change the nature of state-science relations. For the Four Modernizations effort to succeed, the regime needed a state of more-or-less "happy convergence."

Nevertheless the ultimate objective of these reforms was not to redefine state-science relations; it was to reinvigorate the Chinese economy. The Chinese leadership's decision to grant more autonomy to scientists and to institute peer review does not appear to be part of a coherent set of values, comparable to those in the United States or Europe, about how science should be conducted. The decision was motivated instead by near-term economic requirements. The inherent internal contradiction between the scientific ethic and Marxist-Leninist–Mao Zedong thought was not resolved. The Chinese leadership made a trade-off, substituting a short-term risk (give scientists a bigger role) for a long-term benefit (ensure Chinese scientific and technical independence and ultimate superiority). The leadership was hoping to bite off one piece of science reform without swallowing the entire pie. However scientific cultures and values constitute an integrated whole. Thus far it has proved impossible to give scientists a role in decision making and a reasonable degree of autonomy without also allowing them to at least articulate, if not promote, policies that could ultimately threaten the socialist structure of China's economy and society.

The organization and funding of military research deserves special attention for several reasons. First, China's nuclear weapons program is the only post-1949 S&T success story. It has been a source of puzzlement to foreigners that over thirty years ago the Chinese were able to develop, with minimal foreign assistance, an independent nuclear weapons capability, yet are still not able, in 1991, to mass manufacture any of a number of much less sophisticated industrial and consumer products. Second, the entire science system was set up to support military objectives. It was only in the mid-1960s, after the Ministries of Machine Building had expanded their research capability to meet their needs, that any of the elite science institutions began to change their orientation to civilian projects. Most civilian science research has been a spin-off of military projects. Third, the management of military research and development has been quite different from that of civilian programs.

In the nuclear weapons program, for example, it is now known that scientists were given important decision-making roles, they enjoyed a high degree of autonomy, and they were at least buffeted, if not protected, from political and ideological struggle campaigns.[9] A leading rocket scientist described recently how it was possible to make such rapid gains in nuclear science at a time when China was isolated from the rest of the world and lacking in scientific talent.

> Leadership was provided by the Central Special Committee, which was chaired by Zhou Enlai. Nie Rongzhen was in charge of day-to-day affairs. This unified leadership was very powerful and effective. . . . Each project had a technical leader . . . (who) was totally responsible for the technical issues regarding his project. . . The technical organization was very tight. . . . Everyone did a perfect job from the unified leadership provided by the Central Special Committee to the technical and organizational coordination by each individual task force. . . . We already made it happen some 20 years ago. Why can't we accomplish it today?[10]

Information about how Chinese military research is performed, assigned, and evaluated has been difficult to obtain. Most military research is performed in institutes of the CAS and in units reporting to the Ministry of Machine Building and Electronics (MMBE) (see fig. 1), although these organizations also work on civilian projects. Based on the nuclear example, it is safe to assume that state-science relations differ in military and civilian research environments, although how they differ is still a matter for speculation. The Chinese themselves have said that over the past thirty years military research institutes have had access to better manpower, more sophisticated equipment, and more generous funding.[11] However, from what has been determined empirically, these remarks seem to pertain to nuclear research rather than to all Chinese military R&D. The nuclear program appears to have been a separate cell, an isolated complex of institutes and factories, at least somewhat protected from the political upheaval and bureaucratic inertia that has stymied the rest of the science system in China.

---

9. John Lewis and Xue Litai, *China Builds the Bomb* (Stanford: Stanford University Press, 1988), 219–49.

10. Qian Xuesen, "Establishing a Socialist Science and Technology Systems," *Keji Ribao,* September 29, 1989, 1, translated in *JPRS Report: China Science and Technology,* CST-90-030, December 10, 1990, 11–13.

11. See for example, K. C. Yeh, "Industrial Innovation in China with Special to the Metallurgical Industry," Rand Note N-2307 (Santa Monica, CA: Rand Corp.), May 1985; and Jonathon Pollack, "The Chinese Electronics Industry in Transition," Rand Note N-2306, May 1985, 64.

Trade-offs between Military and Civilian Research

The degree to which military priorities should dictate a nation's science agenda is a question of intense interest to the Chinese. As mentioned previously, China's science system was established with the objective of developing nuclear weapons, and therefore favored branches of science and engineering directly related to that (and other military) missions: aeronautical engineering, theoretical physics, materials science, and the like. To that extent, the Chinese responded to security concerns in a way comparable to other sovereign nations: by striving for self sufficiency and scientific autarky in areas of science related to military power.[12]

There are no statistics available to document what percent of the overall R&D budget has been spent on military program, or what percent of the defense budget is dedicated to R&D. There is some evidence to suggest that military R&D declined during the Cultural Revolution, due to the ideological liability connected with any kind of intellectual pursuit during that period. But some programs, notably nuclear research, appear to have been protected, and military spending has not, since 1949, been as high as 2.1 percent of the GNP, and never lower than 1.6 percent of the GNP.[13] Therefore it is no surprise that in China the main scientific achievements since 1949 have been closely connected to military efforts. The fields in which Chinese science has maintained a core of qualified personnel (high energy physics, applied mathematics, aeronautics, quantum mechanics) are all fields that have identifiable military applications. It also follows that China's top scientific talent is now found in institutions closely connected, either now or at some time in the past, to military R&D.

In the early 1980s the Chinese leadership issued numerous policy pronouncements suggesting that resources should be reallocated away from the military sector toward generic, commercial technology that would generate near-term economic returns. Much was made of the fact that national defense was fourth on the list of national modernization projects. The leadership also called for the "demilitarization" of a large number of institutes, schools, factories, research institutes, and personnel, parallelling the significant decline in the size of the standing army and other elements of the military establishment. But it now seems evident that "demilitarization" was desirable only for low-end military research, development, and production facilities, not across the board.

12. Solingen in this volume.

13. Defense spending and GDP/GNP figures from *International Military Balance, 1990–1991* (London: 1991) International Institute for Strategic Studies.

Thus China's reallocation of resources within the science budget was not part of an international trend toward lower defense budgets in Europe, the United States, and the Soviet Union. It was motivated by economic necessity and the impossibility of upgrading the entire military S&T establishment at once.

Whereas the United States and the Soviet Union might decide to refocus research expenditures because they have "overbuilt" their defense capabilities, China's army uses obsolete equipment that badly needs modernization. Thus Chinese policy decisions are possible only because of what Chinese analysts refer to as the "current peaceful international environment," which suggests that it will be a long time before China needs an indigenous capability to build modern weapons. Here it is difficult to tell the chicken from the egg. Do the Chinese really see the environment as more peaceful, or do they use that assessment to justify their desire to develop a nonmilitary scientific base?

Basic versus Applied Science

Generalizations that link attitudes about the relative values of basic and applied research to a particular political structure or type of economy are troublesome. Two reasons for this are the difficulty of empirical international comparisons and the fuzziness of the distinctions between basic and applied science. Scientific projects undertaken without any promise of near-term results, in the interest of national prestige, consistent with a tradition that values the pursuit of knowledge for its own sake, are considered basic science. Before the recent flurry of excitement over superconductivity and cold fusion, both those research endeavors would probably have been considered "basic." Now, because there is a more popular understanding of their potential payoff, they are considered to be applied research. The United States, Japan, and Europe all share certain political and economic structures. Yet each has had a different pattern of allocating resources between basic and applied research. The United States has a history of supporting science projects with no near-term economic or practical payoff. In Japan, on the other hand, applied research has dominated for the past twenty years. Japanese scientists have systematically monitored and assimilated basic research performed elsewhere, primarily the United States and Europe, without themselves engaging heavily in basic research. There is now a concerted effort to bolster basic, conceptual, or, as the Japanese describe it, "creative" scientific work. If the United States begins to shift resources away from basic research in favor of more applied work, or if, because of pressures to redress the competitive balance, Japanese scientists are excluded from U.S. science programs, Japan will not have a conceptual base on which to draw.

For China this issue is especially sensitive for several reasons. Before reviewing the current debate, it might be helpful to summarize what is known about the basic versus applied tradeoff in previous decades. Here again, the available information is scanty. When the post-1949 science system was established, basic research was arguably less strong in applied chemistry, material science, mathematics, and engineering. On the other hand, a country with limited resources and critical economic needs is under pressure to channel those resources to science projects with a near-term payoff. The pressure to catch up, and to catch up quickly, only became more intense with the institution of the "open door" policies of the early 1980s, when it became instantly clear how far China had slipped behind the industrialized world. Thus there is an inherent conflict between China's comparative advantage in science and the immediate needs of the Chinese economy.

Nevertheless, a strong intellectual tradition that values basic theoretical inquiry and the pursuit of knowledge for its own sake persists in China today. Many of the Chinese scientists who have achieved international renown have, in fact, been recognized for their accomplishments in basic theoretical work.

The debate has been clearly resolved in favor of applied science, and that decision has been underscored by funding reforms that provide many more incentives for applied work and might have the result of closing down institutions that do only basic research. The move toward applied research is even more critical when the differences in definition are considered. It might be useful to imagine a spectrum in which basic research is at one end of the continuum and mass production at the other. Visits to many Chinese institutes indicate that much of what the Chinese describe as "basic" research would actually constitute applied research or even development in the United States. China's eighth Five Year Plan included twelve projects all of which are described as "basic" research. These include "high-critical-temperature superconductivity, structure and properties of opto-electronic materials; climate dynamics and climate prediction theory; methods of planning large-scale science and engineering programs; semiconductor superlattice physics, materials and new device structures; male sterile hybrid varieties of grains, cotton, and oil bearing crops; future environmental trends in China; nonlinear science; and crustal movements and global dynamics."[14] Although some of these topics have a theoretical dimension, many would be considered advanced development or engineering in the United States. Thus, the quantity of basic research (as defined by the United States) is actually smaller than the statistics would make it appear.

---

14. "State Projects for Vigorous Developing During Eighth 5-Year Plan Designated," *Keji Ribao*, December 23, 1989, 1, translated in *JPRS: China Science and Technology*, CST-90-07, March 6, 1990, 3.

The funding reforms described above have been structured to promote applied research, since there will be very few sources of competitive funding for theoretical work. Many basic research institutes in China survived decades of political persecution and deprivation only to be subjected to fiscal persecution. If the motivations are different, the results might well be the same. One Chinese scientist wrote, in a letter to the journal *Science* in 1988:

> I think that China's current science and technology (S&T) reform policy, which overemphasizes the commercialization of research results and tries to use administrative measures to coerce scientists into playing the role of businessman, is extremely short-sighted and will do profound damage to the nation's S&T base. . . . The definition of development and the justification of science in purely economic terms are unjustifiably narrow. . . . The new policy has already met with strong opposition from concerned scientists in China, but different opinions have been ignored. . . . Under this policy, research opportunities for both domestic- and overseas-trained scientists and engineers will diminish rather than flourish, because relinquishing the sole responsibility of applied research to industry will result in the neglect of research that has no immediate commercial payoff.[15]

The fact that Japan and the newly industrialized countries (NICs), which have focused on applied research, have vibrant and growing economies, whereas the United States, which has invested heavily in basic research, is experiencing some degree of industrial decline, has clearly been a factor in the Chinese calculus. However China must now ensure that science education focus heavily on applied subjects, and that personnel in abstract, theoretical fields, such as high energy physics, be encouraged or persuaded to shift their research emphasis to more practical problems. Neither task is likely to be easy. In the post-Tiananmen environment, Chinese scientists might well be more concerned about economic survival than about political or democratic reform. Although economic issues might not drive scientists into a deadly encounter with the state, they could well prevent the happy convergence that China arguably needs to sustain in order to move confidently on to the world scene in the next century.

International versus Domestic Science and Technology

In a world where economic interdependence is an unavoidable reality, it seems paradoxical for nations to seek technological protection or independence. Yet

15. *Science* 240 (20 May 1988).

that is precisely what is happening, even as the degree of interdependence is growing. The United States has always sought some form of technological protection for national security reasons. As the fear of technology leakage to military adversaries declines, the fear of giving away U.S. technology to economic competitors is on the rise. The Japanese are also interested in protection of both military and commercial research. However, Japan has managed to ensure access to the international research community without having to share anything of strategic value. Europe has both drawn from and contributed to the international scientific community.

China has undergone periods of isolation, when the leadership has been determined to push ahead with purely indigenous efforts, as well as periods of integration with the international community. In general, the attitude toward foreign science and technology has paralleled broader trends in foreign relations. During periods when China had close alliances (with the Soviet Union, for example), there was more tolerance for large-scale assimilation of foreign technology, know-how, and concepts. Periods of political isolation were accompanied by intense criticism of those who maintained that foreign science was critical to China's emergence as a great power.

The position of the post-Mao leadership has been relatively consistent. Access to foreign science is a near-term requirement, but it must not become a long-term dependence. In this respect, caution is in order when postulating the effect of Chinese foreign economic policies on state-science relations. Although a "liberalizing" economy is often accompanied by "more openness to scientific interdependence," the Chinese leadership is only willing to let interdependence go so far.[16]

Still, the degree of openness that the leadership would tolerate has been quite remarkable. The most dramatic illustration of the leadership's commitment to participation in the international scientific environment was the open door policy that both received foreign scholars and sent Chinese overseas to study. It was inconceivable that China could achieve its national objectives without training a new generation of S&T personnel. Access to foreign universities was critical for two reasons. First, returning students would be able to contribute directly to projects of national scientific and economic significance. Second, and more important, these students would rebuild China's own educational infrastructure, create a set of world class Chinese universities, and thereby ensure technological independence in future generations.

China's decision to send some 100,000 students abroad between 1980 and 1989 was widely hailed as evidence of a long-term and irrevocable commitment to economic and scientific modernization. About 75 percent of the total number of students came to the United States, where their studies were

16. Solingen in this volume.

concentrated in computer science, physics, engineering, math, and biology.[17] As in other areas of modernization, the policy had several unintended consequences. First, exposure to university life in the United States served to highlight for most Chinese students the deficiencies of their own schools, and their own society, materially, intellectually, and politically. The difficulty the Chinese leadership has had placing returned students in positions where they can maintain the same level of professional activity is only one of several reasons why Chinese students in the United States might choose not to return to China. For many, the temptation to stay overseas is quite strong. This contributes to pressure within the leadership to curtail the number of students permitted to study abroad. In addition, study in the United States raised student expectations that change within China would be much more rapid than is probably possible, even under a leadership committed to speedy reform. It was not realistic to expect that Chinese students in the United States would absorb only scientific and technical knowledge and remain hermetically sealed off from cultural influences. Their expectations were transmitted, in person and through international communication channels, back to their classmates in Beijing.

China also runs a risk that access to U.S. or European science will be cut off (or made more difficult) as the result of economic or political disagreements. In fact, the chain of events after the Tiananmen incident illustrated the scientific consequences of unsuccessful reform. Because China's access to the international basic research community was based largely on the goodwill of foreign scientists who wanted to support the free exchange of ideas throughout the intellectual community, that access was also vulnerable to the will of those scientists. The same people who facilitated exchange throughout the 1980s indicated after the Tiananmen crackdown of June 1989 that they are prepared to boycott scientific exchange with the Chinese to express disapproval of Chinese human rights policies. When the United States approved a continuation of Most Favored Nation trade status for China in the spring of 1990, in implicit recognition of political reform in the year following the Tiananmen incident, a group of prominent U.S., Japanese, and European scientists signed a petition stating that they would not attend scientific meetings held in China. The author of the petition, Soviet dissident Yuri Orlov, urged his colleagues to join him in pressuring the Chinese government to release scientists who are political prisoners, citing his experience with the Soviet Union: "It was a long, long process, but the Soviet experience shows that it is not a hopeless task to change their minds."[18] Many individual scientists had already, of their own volition, canceled planned visits to China

17. Leo Orleans, *Chinese Students in America: Politics, Issues, and Numbers* (Chicago: University of Chicago Press, 1973).

18. "Scientists Plan China Boycott," *Science,* 1 June 1990.

in the hope of pressuring the government to ensure intellectual freedom. These individual cancellations, impossible to quantify or document, may ultimately have had more influence on China than formal government sanctions.

Political Activism in the Chinese Science Community

Chinese scientists have not traditionally been at the forefront of political activity. When the party needed their talents (to develop nuclear weapons, for example), they were either coerced or coopted; very few became dedicated activists, either to improve their own professional situation or to promote democratic reform. There have been two specific instances when Chinese scientists have been extremely critical of the Party and of China's S&T system, but in both cases only after the Party specifically asked for their involvement, not because the scientific community demonstrated an interest in becoming a force for political reform. And in both instances they were subject to less persecution than social scientists or humanists.

The first example dates from the 1950s ("Let a Hundred Flowers Bloom") when Mao asked all of China's intellectuals to criticize the Party in an effort to revitalize a dispirited academic community. Scientists, as part of the intellectual community, took part. The criticisms of Party behavior as far back as the Yenan days (1930s) that emerged from this movement were scathing. The result was a classic deadly encounter, in which political accountability replaced most vestiges of scientific autonomy.[19] More than 300,000 intellectuals were branded "rightist" and subsequently isolated, "struggled against," imprisoned, and tortured as a result of their critiques; many promising careers were cut short or postponed for a decade or more.

The second example is the appeal made by the Deng leadership to the scientific community as early as 1977,[20] and repeated that theme until the events of Tiananmen Square in 1989. The Deng regime called out to the scientific community to participate fully in moving China forward to the next century, as part of which scientists were given increased autonomy and international exposure. Although to this case, the scientific community was not

19. Solingen in this volume.

20. The fate of science policy during the Cultural Revolution is not being addressed here because of the difficulty in determining what role members of the scientific community played in that movement and how best to characterize their relationship with the state. There were certainly any number of "deadly encounters," but in the chaos of the movement there was also, during certain periods, some degree of "happy convergence" as well as "passive resistance." In any event, there is nothing to suggest that the scientific community was a prime mover in the Cultural Revolution, either because of its support for specific policies or because of its opposition to Party authority.

explicitly asked for its opinions, the open door and the seemingly relaxed political atmosphere created a climate in which they were more likely to express those opinions. Once again scientists, as well as other intellectuals, came forth with criticisms and suggestions as to how to reform the system to make better use of China's brainpower. Many of the suggestions pertained to improving living conditions for scientists, increasing salaries, permitting some job mobility, reestablishing scientific libraries, encouraging international scientific exchange, sending students overseas for advanced training, and increasing the autonomy and authority of scientists and technical experts. As documented previously, many of these suggestions were taken seriously by the leadership and some were implemented.

Then the terms of the debate were expanded to include questions of epistemology and universal scientific culture as defined by Robert K. Merton.[21] Certain Chinese scientists began to demand, in the name of patriotism, both autonomy and intellectual pluralism. The more outspoken members of the community advocated the overthrow of Marxist dominance of human knowledge. As long as Marxism is at the pinnacle of knowledge, the Party is in a position to choose among the hundred schools of thought and control scientific activities in such a way as to move China toward Communist ideals.

From the earliest days of the Communist party the inherent contradiction between the norms of Western science and the ultimate objectives of the Party were clear to Mao Zedong. Mao had been profoundly uncomfortable with intellectuals and many of the tumultuous political events in post-1949 China were motivated by his inability to resolve this incompatibility. By the mid-1980s the contradiction was reasserting itself in very stark terms, as increasing numbers of scientists were indirectly challenging the Marxist view of science and the pursuit of knowledge.

However, most of the scientific community confined the debate and the challenge to academic journals and theoretical discussions. Only a few scientists had the temerity to advocate specific policies that would promote intellectual pluralism and thereby embody universal scientific ideals. And only one scientist, Fang Lizhi, was willing to go one step further: Fang actually encouraged open defiance of party authority by students as well as faculty. He urged the entire intellectual community to challenge the party head-on on issues that went far beyond autonomy for scientists, or even epistemology. Fang was, in effect, advocating instant democratization of Chinese society on the grounds that without it intellectual endeavors would be meaningless. It is still unclear if his primary preoccupation is the political evolution of Chinese society or the creation of a climate in which scientific inquiry would flourish. If it is the

21. Robert K. Merton, *The Sociology of Science: Theoretical and Empirical Investigations* (Chicago: University of Chicago Press, 1973).

latter, political change constitutes means to an end. Whether one marvels at his courage or dismisses his naivete, it is clear that he is unique among Chinese scientists. The leadership was naturally anxious that Fang not become a symbol or role model; thus the concern over U.S. protection of him at the embassy and the difficulty connected with his ultimate release.

To the extent that there are Chinese scientists who are politically active, they are largely students. Overseas students, many of whom are not going back to China, have played a critical role. Throughout the late 1980s they were able to transmit their enthusiasm about the possibility for change to their classmates in Beijing. Momentum continued to build until there was a standoff with the Party, culminating in the Tiananmen incident of 1989. But the Tiananmen protest quickly broadened in scope to encompass a wide range of demands and dissatisfactions, some of which had only the most indirect connection to scientific issues. Nor does there appear to have been an organized system of political activity on the part of Chinese scientists per se (although they seemed to have supported the students). The early stages of the confrontation were inspired instead by the charisma of a single scientist, Fang Lizhi, who was not necessarily representative of the larger community, and then encouraged by students in the United States who had no intention to return to China.

The Chinese scientific community today, therefore, does not seem to be preoccupied by human rights or socio-economic development. Divergences of opinion with the state are centered on more narrow questions dealing with research priorities, funding for science and technology, and the like. These are, to some extent, "technical" issues, over which people of the same ideological persuasion can legitimately differ. They are not likely to lead to deadly encounters or, in and of themselves, prevent a happy convergence.

Conclusions

Although the available data leave many unanswered questions, one theme emerges: China is a nation searching for a unique model of S&T development, one that will allow it to circumvent some of the difficult tradeoffs between civilian and military science, basic and applied research, and interdependence and autarky in science and technology. Chinese leaders have studied the conduct of science in pluralist market-oriented economies, and they are prepared to adopt some features of that model in the interest of reaping associated benefits.[22] Thus they are more and more open to the involvement of scientists in the decision-making process; they are increasing the use of material incentives; and they are allowing scientists levels of autonomy that would

22. Solingen in this volume.

have been considered unthinkable in earlier decades. However, the leadership continues to resist adopting these reforms in their totality. The scientific community, as a whole, has not been a powerful force for reform. Furthermore, there is a "little bit pregnant" quality about these reforms that makes observers question whether or not they can ultimately succeed in bringing China's considerable S&T resources to bear on current problems and challenges. The leadership wants to take the reform process just far enough to reap near-term economic benefits, but not far enough to run any serious political risk. The difficulty for China is the inherent, inevitable integration of the political economy of science as a coherent whole. This dilemma parallels the tentative, trial-and-error quality of China's economic reforms, which in turn reflects the uncertainty surrounding China's political system in a post-Deng era. Exactly how to organize the nation to face the challenges of the twenty-first century is a subject that the Chinese have been able to address only in the most general terms. Answers are captured in slogans, such as "socialism with Chinese characteristics" or "one country–two systems," that leave most of the difficult questions unanswered. It stands to reason that a nation uncertain about the larger issues of politics, economics, and foreign relations will have a confused and somewhat conflicted approach to state-science relations.

The Political Economy of Soviet Science from Lenin to Gorbachev

Paul R. Josephson

In this century scientists have played an increasingly important role in politics. This is particularly true of industrialized nations where science and technology are vital to the interests of the state. In France, Great Britain, Germany, Japan, the United States, and the Soviet Union scientists give advice, promote policies, and serve the instrumental goals of economic development, health care and welfare, and national security. The contributions of scientists in each of these roles is shaped both by the internal structure of the political economic system and the international context. These factors have an impact on the scale, organization, and funding of science.

The study of the political economy of science in the postwar USSR is one of the most revealing of the impact of internal and external factors on the socioeconomic characteristics of the scientific community, its strategies toward the state, and state strategies toward it. In the 1920s, the state worked closely with scientists to rebuild the scientific enterprise. No coherent national science policy existed. Scientists and bureaucracies sparred among each other for power and influence. Scientists had relative autonomy to determine the specifics of research programs. In the 1930s, indeed until after the death of Stalin in 1953, the state harnessed scientists to the machine of industrialization. The Party brought scientists, like all other potentially autonomous groups, strictly under its control, centralized policy-making, infiltrated research institutes, and established ideological hegemony and international isolation. The government pursued a policy of autarkic scientific relations. It feared hostile capitalist encirclement from without and alleged enemies of the working class within its borders. Heightened ideological scrutiny, coersion, arrest, and execution were employed to ensure compliance with state goals.

After Stalin's death, the Soviet Union set off on a path of economic and political reform that was slowed by the conservatism of the Brezhnev years—which led to the dissolution of the Soviet empire. Under Khrushchev and Brezhnev, scientists succeeded somewhat in loosing the constraints of Party control. Scientists managed to throw off the remnants of the Stalinist legacy only after the revolution initiated by Gorbachev in the 1980s. As in other countries, the increasing importance of science in matters of national security, international prestige, and economic growth led the state to increase support for R&D, especially in such areas as nuclear physics and space research. At the same time, scientific and technological expertise became crucial to the political process in resolving disputes between officials or advancing new programs.

In this chapter, I describe the changing nature of the polity under Soviet leaders from Lenin and Stalin to Gorbachev, and its impact on scientists' political and social responsibilities. I focus on physicists. In spite of divergences in science policy among Soviet leaders, several trends remain constant. First, throughout Soviet history, Party leaders, economic planners, and scientists alike have maintained their faith in modern science as a panacea for the social and economic problems facing their nation. Marxism, which in its Soviet form is particularly economically determinist, has reinforced this belief. Second, scientists and engineers commanded great respect in Soviet history, the periods of ideological scrutiny of their work notwithstanding. And third, the physical and chemical sciences have prospered, while the life sciences have garnered less than their share of resources. Of course, Lysenkoism in genetics contributed to this state of affairs. However, the technicist leaning of Soviet science policy, and the fascination with large-scale technologies tended to reinforce emphasis on the exact sciences.

Stalinism and Science: Revolution from Above

There was no coherent science policy after the Russian Revolution in 1917. Scientific research institutes were found under the jurisdiction of one of a number of different organizations, although primarily under Glavnauka, the Main Scientific Administration of the Commissariat of Enlightenment, and the Scientific Technical Department of the Supreme Economic Council. These two organizations competed for influence in matters of science policy to the benefit of scientists. Scientists played the two organizations against each other, securing funding from both. In a celebrated case, the Leningrad Physico-Technical Institute, the leading physics center in the USSR before World War II, which received money primarily from Glavnauka, reorganized its applied research using the same personnel, equipment, and laboratories, to receive money as the Leningrad Physico-Technical Laboratory from the Scientific Technical Department.

With respect to science, the Russian revolution had run its course with the institution of Stalinist policies in the early 1930s. These policies included the centralization of administration, the introduction of long-range planning, and the requirement that research have immediate effect upon production.[1] In the 1930s, the pressure for centralization of science policy was manifested in the transfer of most major physics and chemistry institutes to the Commissariat of Heavy Industry (Narodnyi komissariat tiazheloi promyshlennosti, or Narkomtiazhprom). The government aimed to control the entire scientific enterprise, from the individual researcher to the Commissariat by means of planning on an all-union scale.

From the first days of the revolution a number of Party officials had called for the reorganization of scientific activity along "socialist" lines of planning, collectivism, and avoidance of parallelism. The rapid growth of the scientific establishment and number of scientists in the 1920s, when the Party had yet to develop a coherent, all-union science policy, frustrated attempts to introduce planning, as did the very small number of scientists holding Party membership, especially among natural and exact scientists.

Around 1929, scientists and their institutes were forced to submit annual and five-year plans of research, spelling out targets and products of research. The government was motivated in part by fear of duplication of research effort. Western observers have argued that the marshaling of resources to avoid duplication prevented competition between scientific centers and made Soviet science less dynamic in many fields than Western science. Initially, scientists presented detailed documents, running hundreds of pages. They quickly discovered, however, that these documents tied their hands, making it difficult to move into new areas of research, or to justify failure to meet goals. They were pleased to learn that central planning agencies were willing to accept more general plans of forty to sixty pages. This, and the largesse bestowed upon scientists as part of the industrialization effort, enabled them to strike out into new areas as the horizons of science shifted, for example, into nuclear physics in 1932. Because of its importance for industrialization, Narkomtiazhprom received massive infusions of funding and grew quite rapidly, both in terms of institutions and scientific workers.

The creation of centralized organs of control was followed by the subjugation of autonomous professional societies. Until the early 1930s, scientists maintained professional organizations in all fields. Chemists, biologists, ecologists, physicists, writers, and architects lobbied the state, held annual or biannual meetings, published proceedings and independent journals. The Russian Association of Physicists (RAF), founded in 1919, had met eight times by 1931 in increasingly well-attended conferences. Under Stalin, however, the party eliminated all professional associations. It feared what it

1. On industrial R&D and technology policy, see Robert Lewis, *Science and Industrialization in the USSR* (New York: Holmes and Meier, 1979); and Bruce Parrott, *Politics and Technology in the Soviet Union* (Cambridge: MIT Press, 1983).

viewed as technocratic aspirations among specialists.[2] In 1931, the RAF was disbanded. The Party created in its place a physics association within Narkomtiazhprom devoted to Stalinist development programs.

Economic planners and Party officials required that research have applicability for the ongoing industrialization effort. This changed the face of science, too. Scientists had to avoid giving them the impression that their research lacked application, or resembled the "ivory tower reasoning" so common in capitalist societies that was divorced from the needs of "socialist reconstruction." Throughout the media, including their own professional journals, scientists were exhorted to force the achievements of science into the productive process in such wide-ranging fields as agriculture, communications, electrification, and metallurgy. In the Academy of Sciences, the creation of a division of technological sciences in 1931, left no doubt as to the intended preeminence of applied science and technology for the Party. Fundamental science lost its hallowed place in Soviet society. The emphasis on applications limited the effectiveness of basic research to this day.

The centralization of science policy was accompanied by the imposition of autarkic scientific relations. Just as in the economy, so in science, scholars were encouraged to go it alone. Scientists had with great difficulty overcome international isolation after World War I and the revolution. Only by 1924 were they regularly traveling to international conferences and studying abroad, receiving foreign publications, and publishing in the west. In the 1930s, these great achievements were abandoned. Regular scientific contacts with the west ceased. Such leading specialists as the biologist Dobzhansky and the physicist Gamov failed to return from western sojourns. The physicist Peter Kapitsa, who had worked for thirteen years in Cambridge, was not permitted in 1934 to return to England after his customary summer holiday in Russia. From this point until the late 1950s, as a rule, only "Party scientists" were allowed to travel abroad. There was even danger associated with sending reprints to foreign colleagues; collaboration with the enemy might be alleged.

It may indeed be that the Party and scientists shared certain goals during the institution of the five-year plans. For example, both wished to expand the research enterprise and to speed up the introduction of the achievements of science into production. Both recognized that government support was vital for increasingly expensive, large scale research programs in modern scientific institutions. Scientists worried about the encroachment on their enterprise by the state. They feared premature identification of targets, arguing that it was often impossible to identify future fruitful areas of research. However, because of continued achievements in nuclear, solid state and theoretical

---

2. Kendall Bailes, "The Politics of Technology: Stalin and Technocratic Thinking Among Soviet Engineers," *American Historical Review* 79 (1974): 445–69.

physics, and in the applied fields of communications, heat engineering, and electrification, they were able to avoid some of the more onerous restrictions placed on them by the introduction of Stalinist policies.

Stalinism and Science: Cultural Revolution from Below

Cultural revolution accompanied the Stalinist revolution from above.[3] An increasingly proletarian Communist party demanded class war against privilege, and against bourgeois specialists; that is, those who had received their education during the Tsarist regime or whose parents were from the intelligentsia. Cultural revolution was intended to lead to the replacement of bourgeois specialists with scientists of the proper social origin and worldview. This had far-reaching impact upon basic science in the areas of the relationship between scientist and Party, the philosophy of science, and international relations. The Party signaled its intention to limit the autonomy of specialists and mobilize them for the industrialization effort through the Shakhty and Promparty Affairs in the late 1920s. These highly publicized show trials revealed that scientists and engineers were "wrecking" state industrialization plans.

Cultural revolution included attempts to take over the administration of scientific research institutes through the penetration of communist cadres into the ranks of scientists through cooptation, coercion, and *vydvizhenie*. *Vydvizhenie* was the advancement of workers into positions of administrative responsibility in economic enterprises, higher educational institutions, and scientific research institutes on the basis of class origin and Party affiliation rather than on merit, qualifications, or other traditional reasons for advancement.

Whole organizations and institutes were subjected to the Party assault, with varying degrees of success. These included the Academy of Sciences. The Party had discovered through a series of investigations in the late 1920s that most fundamental research institutes lacked sufficient Communist personnel. Leningrad was the center of scientific activity with the Academy of Sciences, and other leading research institutes, but in 1929, of over 5,000 scientific workers in the city, only 39 were Party members.[4] Of 25,286 "scientific workers" in the USSR surveyed in 1930, only 2,007 (less than 8 percent)

3. For a discussion of some of the issues in this section, see Paul R. Josephson, "Soviet Scientists and the State: Politics, Ideology, and Fundamental Research from Stalin to Gorbachev," *Social Research* 59, no. 3 (Fall 1992): 589–614. This special issue of *Social Research,* under the editorship of Margaret Jacob, is devoted to science and politics.

4. *Partrabotnik* 15 (1929): 53, as cited in K. E. Pechkurova, *Partiinoe rukovodstvo nauchnykh kadrov v gody pervoi piatiletki (1928–1932 gg). (na materialakh Leningrada)* (Candidate's dissertation, Leningrad State University, 1976), 6.

claimed Party membership, and of these only forty-four were physicists, with but three in Leningrad. Two-thirds were social scientists; only 8 percent worked in the exact sciences.[5] Essentially, the process of infiltration involved the elevation of the role of the primary party organization within research centers to one of close scrutiny of the scientific directors. This ensured that institutes followed instructions issued from above. The process was often coercive and usually involuntary from the perspective of the institutions.[6]

The Party was less than successful in its attempts to bring more communists into scientific research institutes, or to train new communist scientific cadres. This was especially true of the natural and exact sciences, and particularly among world class scholars. Most leading Soviet scientists joined the Party after some delay, late in their careers. Obligatory Party membership for institute directors is largely a feature of the post-Stalin era. The major methods of increasing Party membership in scientific institutions were *aspirantura* (graduate training), with candidates having a better chance of admission if they had Party affiliation and working class roots; *vydvizhenie,* and Marxist study circles. Initially, there was success in quantity, but little in quality. The *vydvizhentsy* often lacked rudimentary skills, it was difficult to train them, and in fact there were very few "worker-scientists." *Vydvizhenie* was scaled back in 1934 when it became clear that advanced workers were more trouble than they were worth.

In addition, to *vydvizhenie* and *aspirantura,* there were two mechanisms by which the Party "infiltrated" scientific organizations. One was conversion and cooptation, the other was the coercive methods of the purges. Conversion and cooptation involved setting up study circles on the history of the Communist Party, dialectical materialist philosophy, and (scientific) methodology. This approach seems to have been adopted in most scientific research institutes. Many of these circles operated until the late 1970s.

Between 1936 and 1938, the scientific community fell into the maelstrom of the Great Terror. At least eight million people perished in the purges. Thousands of scientists were arrested, many served in the Stalinist gulag, and a large number were shot. From the future Nobel prize-winning physicist Lev Landau, who served a year in prison, to the leading geneticist Nikolai Vavilov, who perished in the camps, but whose brother, Sergei, became president of the Academy of Sciences immediately after the Great Terror, seemingly no one was immune.

---

5. G. Krovitskii and B. Revskii, *Nauchnye kadry VKP (b)* (Moscow, 1930), 16, 51–83.

6. See Loren Graham, *The Soviet Academy of Sciences and the Communist Party, 1927–1932* (Princeton: Princeton University Press, 1967); and Paul R. Josephson, *Physics and Politics in Revolutionary Russia* (Berkeley, Los Angeles, Oxford: University of California Press, 1991).

The Ideologization of Science

The result of cultural revolution and the Stalinist revolution from above was the ideologization of science. From the late 1920s until the early 1960s, philosophers and ideologues drew sharp distinctions between Soviet and capitalist science. They insisted that scientists recognize that their research activities were inherently political. Assisted by Marxist scientists they argued that certain fields of modern science—genetics, relativity theory, and quantum mechanics, for example—were idealist, or pseudoscience, and hence harmful to the proletariat. Entire fields were proscribed, institutes closed, scientists arrested and shot.

The ideologization of science was based partially upon the resurrection of the attitude toward science of the proletarian culture movement (1917–21). In addition to a mistrustful attitude toward "bourgeois specialists," those who embraced proletarian science believed it was based on the principle of "collective" scientific activity, and on planning to avoid the duplication endemic in bourgeois science. Proletkultists contended that a new methodological approach was needed and "pure" science—i.e., science for its own sake—should be prevented. The Proletkultists believed that the revolution that would create socialist "productive forces" and "productive relations," would have to be accompanied by proletarian institutions, proletarian philosophy, and proletarian science. The proletarian culture movement did not long survive the revolution. The party objected to its rejection of pragmatic cooperation with "bourgeois specialists." However, belief in the existence of "proletarian science" persisted. This contributed to a series of fateful decisions to regulate the content and methodology of scientific activity.

In the late 1920s and 1930s, an increasingly stormy debate among Marxist philosophers over the relationship between the Soviet philosophy, dialectical materialism, and science spilled over into the scientific community at large. The participants disagreed over how to apply the teachings of Marx, Engels, Lenin, and ultimately Stalin, to the form and content of modern science.[7] The resolution of the debate was that Stalinist ideologues acquired the authority to inform scientists which approaches were acceptable in the proletarian USSR. In genetics, the result was the rejection of the gene. In 1948, at a national conference at the All-Union Lenin Academy of Agricultural Sciences, Trofim Lysenko, hack-scientist, was proclaimed victor, genetics was outlawed and eliminated from textbooks and universities until 1965.

---

7. On these philosophical debates, see David Joravsky, *Soviet Marxism and Natural Science, 1917–1931* (New York: Columbia University Press, 1961); Loren Graham, *Science Philosophy, and Human Behavior in the Soviet Union* (New York: Columbia University Press, 1987), and Josephson, *Physics and Politics,* 213–75.

In physics, the results were nearly as severe. For twenty years ideologues and party philosophers had attacked "mathematical formalism," the Copenhagen interpretation in quantum mechanics, and relativity theory. Physicists escaped the wholesale debilitation that hit biology and genetics under Lysenko for a number of reasons. Through participation in Marxist study circles they were well prepared to discuss epistemological issues from a position of strength. The complexity of philosophical issues surrounding relativity, quantum mechanics, and nuclear and solid state physics made it difficult for ideologues to challenge them. And they presented a united front to ideological attacks on their discipline.

In spite of their accomplishments for the state concerning the economy in the 1930s and in the recovery from World War II, physicists nonetheless faced increasing scrutiny for alleged idealism, particularly the Copenhagen interpretation of quantum mechanics during the Zhdanovshchina.[8] The Zhdanovshchina, an attack on "cosmopolitanism" and "idealism" in Soviet society, began in 1947. The Zhdanovshchina involved unrelenting philosophical pressure on scientists and artists to conform to the Stalinist standard and to avoid "kow-towing" before Western ideas. In 1948, a series of meetings was held to condemn the alleged idealism rampant in modern physics. A number of physicists working on the atomic bomb project got wind of the idea, called Beria, head of the secret police, and informed him a bomb could not be constructed without taking note of relativity theory and the equivalence of matter and energy. The conference was not convened, but there was a stultifying impact on theoretical physics.

The unrelenting ideological pressure on physicists culminated in the publication of a collection of articles in 1952 that subjected leading representatives of quantum mechanics and relativity theory at home and abroad to intense scrutiny and criticism. No one was immune. A. F. Ioffe, dean of Soviet physicists, came under attack and was removed from directorship of the Leningrad Physico-Technical Institute. Ioffe, a specialist in solid state physics and leading organizer of the Russian Association of Physicists in postrevolutionary Russia, wrote widely in the philosophy of physics. In the late 1930s he had passionately defended Soviet physicists from Stalinist ideologues and such physicists as V. F. Mitkevich and A. K. Timiriatsev from charges that they were idealist for embracing action at a distance and rejecting a mechanistic, antirelativistic ether. He accused his detractors of being reactionary, antisemitic, and of having anachronistic conceptions of physics. He helped other leading physicists defend physics from charges of idealism.[9] In

8. See Werner Hahn, *Postwar Soviet Politics: The Fall of Zhdanov and the Defeat of Moderation, 1946–53* (Ithaca: Cornell University Press, 1982); and Graham, *Science, Philosophy, and Human Behavior,* chaps. 10 and 11.

9. See A. F. Ioffe, "O polozhenii na filosofskom fronte sovetskoi fiziki," *Pod znamenem marksizma* nos. 11–12 (1937): 133–143.

1949 Ioffe published a major treatise devoted to the defense of contemporary physical conceptions of relativity theory and quantum mechanics that soon came under attack.[10] Ioffe's most complete response to this criticism, and one of the first attempts to reassert physicists' control over the philosophy of physics, was published in 1954 after the death of Stalin in a letter published in the leading theoretical physics journal.[11]

Ioffe's defense of contemporary physics in the last year of Stalin's life was not an isolated incident. During the height of the Zhdanovshchina and into the early 1950s such scholars as L. D. Landau, I. E. Tamm, and V. A. Fock wrote extensively about the compatibility of the new physics with dialectical materialism. Physicists drew strength in the discussions with Stalinist ideologues from their continuing success in such areas as elementary particle, nuclear, and solid state physics. If the correctness of Soviet philosophy was verified in "practice," then the development of atomic weapons, nuclear power engineering, and space technologies indicated just how correct new philosophical conceptions in physics must be. As physicists' research showed success after success, and achievement after achievement, the party came to recognize physics as the "leading science."

Shortly after Stalin's death, physicists reasserted their primacy in matters of philosophy. They became more and more outspoken especially on the eve of the fiftieth anniversary of the theory of relativity. At a special session of the Division of Physico-Mathematical Sciences of the Academy of Sciences on 30 November 1955, Tamm, Landau, V. L. Ginzburg, Fock, A. D. Aleksandrov, and others described the thoroughly materialist nature of Einstein's work.[12] Over the next several months physicists and philosophers of science published two dozen articles that adopted the same position in all the leading scientific journals. In the summer of 1956, scientists called for a conference to establish their priority in matters of philosophy of science and philosophy, and to criticize "improper" philosophy of physics.[13] This led to a special all-union conference of the Academy of Sciences to discuss the philosophical problems of the natural sciences in October 1958. The participants included the presi-

10. See N. F. Ovchinnikov, "Massa i energiia," *Priroda* 11 (1951): 7–16; "'Fizicheskii' idealizm—vrag nauki," *Nauka i zhizn'* 3 (1952):42–46; and I. V. Kuznetsov, "Lenin i estestvoznanie," *Nauka i zhizn'* 1 (1951): 1–6. See also M. E. Omel'ianovskii's review of Ioffe's book in *Voprosy filosofii* 2 (1951): 203–7; and *Uspekhi fizicheskikh nauk* (1951–52) for a series of articles. See also Alexander Vucinich, *Empire of Knowledge* (Berkeley, Los Angeles, Oxford: University of California Press 1984), 222–24; and Graham, *Science, Philosophy, and Human Behavior in the Soviet Union,* 355–66.

11. A. F. Ioffe, "K voprosu o filosofskikh oshibkakh moei knigi 'osnovnye problemy sovremennoi fiziki'," *Uspekhi fizicheskikh nauk* 52, no. 4 (August 1954): 589–98.

12. N. a., "Piatidesiatiletie teorii otnositel'nosti," *Priroda* 1 (1956): 114; and V. N. Lazukin, "50-letie teorii otnositel'nosti," *Vestnik Akademii Nauk SSSR* 2 (1956): 106–10.

13. M. E. Omel'ianovskii, "Zadachi razrabotki problemy 'dialekticheskogo materializma i sovremennogo estestvoznaniia'," *Vestnik Akademii Nauk SSSR* 10 (1956): 3–11.

dent of the Academy and leading representatives from all fields. The participants asserted that they, and not Stalinist ideologues or policymakers, should resolve issues in the philosophy of physics (relativity theory, quantum mechanics, motion), cybernetics, cosmology, biology, the origin of life, and so on. For their part, physicists instructed those who had concluded that modern physics was idealist due to the philosophical stands of some of its Western representatives to recognize that its physical content surely confirmed dialectical materialism.[14]

Another basis for physicists' ability to counter Stalinist science policy, and for their prestige in postwar Soviet society was their role in the victory over Nazi Germany.[15] In the glow of their successes, and in concert with other scientists, physicists hoped to use the hard-fought and costly victory over Nazi Germany to support their calls for greater autonomy for their scientific institutes. In fact, the sciences rapidly received increased support to help rebuild Russia's devastated economy in the mid- to late-1940s. Starting with the celebration of the Academy's 220th anniversary in 1945, scientists pushed for more regular ties with Western scholars, greater control over the development of their disciplines, including the ideologically suspect genetics and quantum mechanics, and larger budgetary appropriations. However, in the postwar years, once again the Party attempted to ensure the central direction of science policy through the formation of the State Committee for the Introduction of New Technology into the Economy of the USSR (GKVNT, founded in 1947 by central committee decree). This gave way to the State Committee for the Coordination of Scientific Research (GKKNIR, founded in 1961), and finally to the State Committee for Science and Technology (*Gosurdarstvennyi kommitet po nauke i tekhnike,* or GKNT) in 1965 "to supervise and regulate scientific research, development and technical innovation."[16]

The Gigantomania of High Stalinism

In addition to the centralization of policy-making and the ideologization of science, Stalin placed his signature on science and technology through hom-

14. The papers for the 1958 conference have been collected in P. N. Fedoseev, ed., *Filosofskie problemy sovremennogo estestvoznania* (Moscow: Akademii nauk SSSR, 1959). For a discussion of the 1958 conference and its results see: Alexander Vucinich, *Empire of Knowledge* (Berkeley: University of California Press, 1984), 330–35; *Vestnik Akademii Nauk SSSR* 1 (1959): 132–38; *Voprosy filosofii* 2 (1959): 67–84; *Nauchnye doklady vysshoi shkoly. Filosofskie nauki* 4 (1958); 218–21; and *Vestnik vysshoi shkoly* 2 (1959): 39–47.

15. G. L. Sobolev, *Uchenye Leningrada v gody velikoi otechestvennoi voiny, 1941–45* (Moscow-Leningrad: Nauka, 1966). See also, A. A. Baikov, "Sovetskaia nauka za tri goda voiny," *Vestnik akademii nauk SSSR* 7–8 (1944): 18–22; and A. P. Grinberg and V. Ia. Frenkel', *Igor' Vasil'evich Kurchatov v fiziko-tekhnicheskom institute* (Leningrad: Nauka, 1984), 137–45.

16. Louvan Nolting, *The Structure and Function of the USSR State Committee for Science and Technology,* Foreign Economic Report, no. 16, U.S. Department of Commerce, Bureau of the Census (Washington, D.C.: GPD, 1979), 1–8.

age to projects of gradiose stature. The view of science and technology as a panacea central to Soviet Marxism dates to the first days of the revolution. G. A. Cohen argues that historical materialism is a technologically determinist doctrine. This means that the development of the productive forces (machines, tools, instruments, science, technology, and people themselves) is the single most important factor in the course of human history, and, further, that science and technology provide the key to the socialist future.[17] Some have disputed Cohen's contention that for Marx "machines make history."[18] But whether Marx so argued does not obscure the fact that Soviet theorists and political leaders from Lenin on have embraced this view. They see technology as "the highest form of culture," emphasize the development of the productive forces in the creation of communism, declare that "technology decides everything," and put their faith in science to achieve political, economic, and social goals.

Especially in the postwar years of the atomic and hydrogen bomb projects, nuclear power, and rocket development, faith in technology—and its practitioners, the physicists—did not abate. It was central to the construction of Communism. Its display value—not merely physical presence, but ideological significance—grew at the Magnitogorsk and Norilsk steel combines, the Dneprostroi hydropower station, the Stalinist skyscraper, the construction of the metro and canals, and other projects infamous for their gigantomania, if not aesthetics. As Douglas Weiner tells us, even the environment was seen as something to be tamed, subjugated to the planner's will.[19]

In the late 1940s and 1950s, a cult of science, a logical continuation of Stalinist gigantomania, put scientists and engineers at the forefront of the effort to use big science and technology to solve Soviet economic problems. The cult grew unchallenged in the 1950s on the foundation of successes in space, nuclear power, and high energy physics that rivaled those in the West.[20] The cult was part of the general environment of de-Stalinization during which Soviet scholars, and in particular physicists, began to reclaim control over the scientific enterprise. The cult of science was based on an attitude that such large-scale, expensive, and visible projects as nuclear fission and fusion reactors, high voltage, long distance power lines, particle accelerators, and space ships should be at the center of R&D programs for national, international, and economic reasons. While some public displeasure

---

17. G. A. Cohen, *Karl Marx's Theory of History: A Defence* (Princeton: Princeton University Press, 1978).

18. Donald MacKenzie, "Marx and the Machine," *Technology and Culture* 25 (1984): 473–502.

19. Douglas Weiner, *Models of Nature: Ecology, Conservation and Cultural Revolution in Soviet Russia* (Bloomington: Indiana University Press, 1988).

20. Paul Josephson, "Rockets, Reactors and Soviet Culture," in Loren Graham, ed., *Science and the Soviet Social Order* (Cambridge: Harvard University Press, 1990), 168–91.

with expenditures on technologies with great display value but limited significance for the consumer can be identified, by and large this opposition was muted until Gorbachev came to power, when glasnost and perestroika triggered a reevaluation of the place of science and technology in society.

On the eve of the Twentieth Party Congress in 1956, which is known primarily for Khrushchev's secret speech condemning the excesses of Stalinism, Soviet physicists had achieved substantial institutional and technological momentum. Their R&D programs in space and nuclear science had entered the deployment stage; their victory over Stalinist ideologues in the philosophy of physics had been assured by their consistent defense of the discipline; and the cult of science had given a boost to both their political and programmatic aspirations. Physicists dominated such national scientific institutions as the Academy of Sciences in terms of numbers of institutes and positions of authority from which they were able to exert their influence since the early 1930s. Scientists, government officials, and journalists joined together to stress the central place of physics in plans for economic development and reconstruction of the USSR, especially in the areas of automation, communications, telemechanics, and atomic energy.[21] As the leaders within the cult of science, physicists found the political, social, and cultural bases with which to reestablish control over the institutions of their discipline, offer refuge to geneticists from Lysenkoism within their institutes,[22] and advance more and more fantastic visions of a future Communist society based on the achievements of science and technology.[23]

---

21. For example, A. N. Nesmianov, "Nekotorye problemy sovetskoi nauki, *Vestnik Akademii Nauk SSSR* 5 (1954): 3–25, and "Ob osnovykh napravleniiakh v rabote AN SSSR," *Vestnik Akademii Nauk SSSR* 2 (1957): 3–42; L. L. Miasnikov, "Lider sovremennogo estestvoznaniia," *Nauka i zhizn'* 4 (1955): 10–12; Oleg Pisarzhevskii, "Fizika i tekhnika," *Fizika v shkole* 6 (1957): 5–15.

22. Mark Adams, Zhores Medvedev, and others have written about the protection afforded some geneticists within physico-mathematical institutes. See Mark Adams, "Science, Ideology and Structure: The Kol'tsov Institute, 1900–1970," in Linda Lubrano, Susan Solomon, eds., *The Social Context of Soviet Science* (Boulder: Westview, 1980), 173–205; and Zhores Medvedev, *The Rise and Fall of T. D. Lysenko* (New York: Columbia University Press, 1969).

The role of atomic energy in freeing biologists to study genetics is rather important here. The study of the effects of radioisotopes and radioactivity on living organisms (for example, as medical treatment, as a cause for mutation, etc.) led to increased interest in the study of genetics and the mechanisms of inheritance. See A. N. Nesmianov, "O zadachakh AN SSSR v svete XX s"ezda KPSS," *Vestnik Akademii Nauk SSSR* 6 (1956): 5–6; and N. P. Dubinin, "Problemy i zadachi radiotsionnoi genetiki," *Vestnik Akademii Nauk SSSR* 8 (1956): 22–33.

23. One aspect of the cult of science was the "cult of the atom," based on both reasonable and somewhat far-fetched visions of nuclear energy's potential. The uses ranged from radioisotopes in industry, agriculture, medicine, and biology to nuclear powered automobiles, locomotives, airplanes, and rockets, and finally a Soviet version of "Project Plowshares," geographic and geological reconstruction of nature with thermonuclear devices. All of these visions rested on the assumption that atomic energy would accelerate the construction of communism. See Paul R.

Scientists and Khrushchev

Khrushchev hoped to use Soviet scientific and technological successes to bolster his position within the Party. He embraced large-scale R&D programs in nuclear and high energy physics, chemistry and agriculture (the planting of corn, sowing of Black Earth regions, the wide-spread application of fertilizers and pesticides). He presided over the rapid expansion of the scientific enterprise by all sorts of quantitative measures: numbers of scientists, institutes, and publications. He advocated the creation of a series of science cities in Novosibirsk (Akademgorodok), and around Moscow (Dubna, Khimki, Chernogolovka, Pushchino, and many others). Finally, as part of the de-Stalinization thaw, Khrushchev supported increased autonomy for scientists in the day-to-day management of their enterprise. His attitude toward science as a panacea was symbolically revealed in his visit to Harwell, England, in April 1956, accompanied by I. V. Kurchatov, to see the major British nuclear research facility, and his participation in the Geneva Conference on the Peaceful Uses of Atomic Energy.[24]

Of course, the most important scientific achievements of Khrushchev's years occurred in space technology. Like those in nuclear power, these were key components of Soviet foreign policy. As in the West, millions were spent on programs with limited social utility and questionable technical feasibility. Potential military applications justified expenditures. The absence of public access to information about the programs contributed to the technological arrogance that dominated program decision making. These programs include a "project plowshares" to develop thermonuclear devices for geological engineering, a nuclear airplane like the U.S. ANP, and a series of satellites with nuclear reactor power sources. The significance of space and nuclear research for Soviet prestige was surpassed only by their military significance. These successes were the capstone of physicists' increasing power and influence under Khrushchev. A series of "firsts"—the first satellite, man in space, two-manned shot, woman in space, space walk, soft landing, and so on—convinced the Soviet populace, if not the majority of party officials, of the superiority of the Soviet science. Khrushchev and other officials knew of the technological failings, if not backwardness of their program, and knew that

---

Josephson, "Atomic Energy and 'Atomic Culture' in the USSR: The Ideological Roots of Economic and Safety Problems Facing the Nuclear Power Industry After Chernobyl," in T. Anthony Jones, David Powell, and Walter Connor, eds., *Soviet Social Problems* (Boulder: Westview Press, 1991), 55–77.

24. See N. S. Khrushchev, *Khrushchev Remembers,* trans. Strobe Talbott (Boston: Little, Brown and Co., 1974), 58–71, for a discussion of Khrushchev's relationship with the scientific intelligentsia. See *Atomnaia energiia* 1, no. 3 (1956): 6–7, 66–67 for photographs of the Harwell trip.

the United States would respond to Soviet achievements with redoubled efforts in space. Still they embraced Sputnik as confirmation that a policy that gave scientists increased autonomy in the design and administration of R&D was in the best interests of the country.[25] The anomaly of Khrushchev's rule with respect to science concerned biology and genetics. While the late 1950s saw the reemergence of genetics under the protection of mathematics and physics departments and at the newly founded Akademgorodok (Science City) in Novosibirsk, Khrushchev remained entranced by the peasant wiles of Trofim Denisovich Lysenko. Only after Khrushchev's ouster did scientists manage to rid themselves of Lysenko.

Under Khrushchev, scientists reestablished autonomy in several areas of their activity. Leading scholars pushed for increased control over fundamental research. They secured the transfer of technical sciences from the purview of the Academy of Sciences to the industrial ministries in 1961, ending a thirty-year period during which the technological division of the Academy had been foisted upon it by Stalinist planners. Scientists succeeded in weakening the Stalinist precepts of the unity of theory and practice, and in ignoring the lurking danger of idealism. They had taken a leading role in resurrecting constructivist visions of the communist future. They had yet to regain authority to function in professional associations, but were prepared to assume greater political visibility by virtue of their important skills in the areas of economic development and national security.

Soviet scientists were actively engaged from this time as experts in arms control negotiations. At home, they studied verification and confidence-building measures. Many believed it was impossible to cooperate with the United States. They argued the United States was ahead in arms development and capable of deception in treaty monitoring. The majority believed in the necessity of negotiating an arms control agreement. Until the archives have been studied further, it will be difficult to determine the extent to which those who participated in Pugwash conferences and other international regimes were merely pawns of the state. Soviet scientists with whom I have spoken believe that the international activity of Soviet scientists was strictly controlled, and that true discussion of the political and military issues took place at home behind closed doors. The treatment of such dissidents as Andrei Sakharov and Iuri Orlov indicates the extent to which the power of physicists was restricted. Dissent was tolerated only within the system. Those who went public with human rights or arms control concerns lost their jobs and were harrassed or arrested.

---

25. On space culture, see Paul R. Josephson, "Rockets, Reactors and Soviet Culture," in Loren Graham, ed., *Science and the Soviet Social Order,* (Cambridge: Harvard University Press, 1990), 168–91.

The reformist tendencies embraced by both the Party and scientists during the Khrushchev years were intended to improve the efficiency of scientific research and development. During Khrushchev's rule, the rate of growth of numbers of researchers, institutes, publications, and other quantitative indices of performance increased nearly geometrically. The Soviets were among world leaders in fields ranging from elementary particle and nuclear physics to the conquest of space. Unfortunately, the judgments were not nearly as satisfactory from a qualitative perspective. Whether by such indices as Nobel prizes and scientific citations, or by more subjective evaluations as those offered by the Western peers of Soviet scholars, Soviet science did not fare as well. During the Brezhnev years a series of administrative measures was applied unsuccessfully to deal with this problem.

Stagnation of Reform under Brezhnev

Under Brezhnev, reformist tendencies ran out of steam. Brezhnevites attempted to improve the performance of science and technology by administrative fiat. A conservative group of elder statesmen of science whose primary goals were the preservation of their authority and pet research programs, and the advancement of their students into positions of responsibility, dominated scientific policy-making. Brezhnev had criticized Khrushchev's lack of "trust in party cadres" when the latter was ousted; Khrushchev had often gone outside of normal channels to set reforms in motion. In concert with Brezhnev appointees, such leading scientists as N. G. Basov, a founder of quantum electronics, and A. P. Aleksandrov pushed through half-hearted measures of reform. These measures—the formation of national bodies for the coordination of R&D, the creation of scientific-production associations, and incessant calls to embrace the advantages of the developed Socialist state during the ongoing scientific-technological revolution—amounted more to reform on paper than in reality. The result was the ossification of scientific policy-making in the hands of scientific bureaucrats, and a continued deterioration in the performance of Soviet science. The system proved capable of pioneering efforts in space (Sputnik), nuclear fusion (Tokamaks) and fission, and elementary particle and theoretical physics, but could not maintain a lead or catch up in areas where it was behind.

From an ideological standpoint, insidious controls remained. Pressure to conduct research of an applied nature and to accelerate the introduction of the achievements of science into the economy increased. Strict controls on Western scientific literature and contacts remained. Overt discrimination against such national minorities as Jews persisted, especially concerning entry to universities and institutes, or travel abroad. Often researchers were rewarded with travel abroad more on the basis of party affiliation than scientific quality.

Military R&D came to dominate national programs. The series of reforms set forth during the Brezhnev years reflected the belief among party officials and scientific bureaucrats that central planning of R&D and command economic mechanisms should remain in place. The hope was that long-range planning would permit greater flexibility in R&D by tying it less closely to short-term economic needs. Planners also hoped that the creation of so-called scientific-production associations that united research and the production process in one organization would increase the effectiveness of scientific research and accelerate *vnedrenie* (assimilation of the achievements of science and technology in the productive process).[26] Since the associations lacked the ability to set prices and the wherewithal or contractual ability to acquire the goods and services necessary for production, they were doomed to failure. The fascination with economies of scale also contributed to this failure as it became a simple matter to fund nuclear fusion or fission, or space R&D, but not more innovative, initially small-scale projects in such sunrise industries as computers and biotechnology.

Other major handicaps on the performance of science within this system remained. These included the separation of responsibilities for fundamental and applied research, technological development, and training among the Academy of Sciences, industrial ministries, and universities, respectively; a mismatch between human and material resources exacerbated by an overly centralized system of allocation of inputs and outputs; and continued efforts to create a central administrative body to force the pace of research and development.[27]

On the eve of Gorbachev's rise to power, the power and autonomy of scientists was limited by policies that were intended by Party officials to keep decision making in their hands, and by constraints of the economic and ideological systems that favored the development of "big physics." The reformist tendencies of the Khrushchev years were abandoned. Physicists, who had gained authority through the cult of science and managed to free the Academy from pressure to conduct applied research now faced scientific bureaucrats who wished to maintain the status quo in the management, administration, and structure and funding of science. Indeed, some of the bureaucrats were the very same physicists. The Party was determined to control the entire scientific enterprise, from the level of the individual researcher and institute to such national organizations as the Academy of Sciences, and state committees, and placed emphasis on large-scale projects with visible results

26. Louvan Nolting, *The Planning of Research, Development and Administration in the USSR*, Foreign Economic Report, no. 14, U.S. Department of Commerce, Bureau of the Census (Washington, D.C.: GPD, 1978).

27. The best single treatment of the issues and actors of Soviet science policy is Stephen Fortescue, *Science Policy in the Soviet Union* (London and New York: Routledge, 1990).

and military application but less immediate social utility: rockets, reactors, particle accelerators, military hardware, river diversion programs, and so on.

Gorbachev and Scientists

The reforms in administration and funding of science initiated under Gorbachev destroyed what was left of the Stalinist legacy. The Gorbachev reforms entailed decentralization of administration and funding of research, democratization of management from the level of the Academy to the level of the institute, an end to centralized control of information and foreign contacts, and the creation of informed government scientific advisory groups.[28]

The Gorbachev reforms resulted in a new relationship among experts, government, and the public. The Supreme Soviet, the parliament that sat for the final three turbulent years of Gorbachev's rule, had a limited role in science policy. While one-eighth of all deputies of the Supreme Soviet were scientists, they failed to form a cohesive interest group. Instead, all scientists—physicists, biologists, engineers, chemists, and so on—organized new associations both to defend their professional interests and ensure their input in decision making. Local governments began to assert their prerogatives in science and technology policy, having grown concerned about the social and environmental costs of unregulated industries and nuclear power, and adopted a "NIMBY" (not-in-my-backyard) attitude to any technology suspect in the least.

Several changes in the structure, funding, and administration of scientific R&D accompanied perestroika and contributed to scientists' ability to play a prominent role in contemporary politics. These changes included measures intended to relax the hold on scientific R&D of the older scientific bureaucrats, advancing those to the forefront who support Gorbachev; to improve the efficiency of applied research; to cede some of the power of Party organizations to the Academy of Sciences, institute academic councils, and newly formed ad hoc scientific advisory committees; and to speed up the dissemination of scientific information by easing up on censorship, restrictions on foreign travel, and accelerating the use of computers. Changes also included efforts to secure greater cooperation with Western scholars in many fields of research. These measures depended on Gorbachev's support for reform of the economy and political system.

Gorbachev embraced specialists' advice in a way unparalled in Soviet politics. He placed scientists at the forefront of economic reforms. He surrounded himself with a group of leading academic specialists who served as

28. Paul R. Josephson, "Scientists, the Public and the Party Under Gorbachev," *Harriman Institute Forum* 3, no. 5 (May 1990), 1–8.

an informal science advisory group. Such physicists as E. P. Velikhov and Zh. I. Alferov, and the economist A. G. Aganbegyan (who was initially a considerable influence in economic planning), and others took leading roles to ensure the success of the ongoing political and economic reforms. They stressed the importance of independent scientific expertise in decision making. The increasing visibility of science advisors had significant benefits for society in terms of invigorating R&D, raising awareness of environmental degradation, and modernizing the Soviet economy. The last area was particularly vital; largely because of bureaucratic impediments to innovation the USSR lagged far behind the U.S., Europe, and Japan in such sunrise industries as biotechnology, fiber obtics, superconductivity, and computers. Through the leadership of scientific specialists the Soviets hoped to revitalize these industries.

From 1988 onward, the Party and government attempted to accelerate scientific progress through reforms of the economy. These involved the first steps toward restoration of market mechanisms in limited sectors of the economy. One of the most promising ways to increase the efficiency of research was to introduce *khozraschët* (self-financing) into the scientific enterprise. This required institutes to do more research under contract with a variety of bureaucracies and economic enterprises, and to secure grants from government and scientific societies in addition to their normal budgets. Those that failed to do so profitably would fall by the wayside. In addition, in the past it was impossible to fire personnel in most cases. Once an institute was created it was unheard of to close it down or change its focus even when it had outlived its usefulness. The effort to make institutes more cost-effective, by ending both the hoarding of workers and underemployment, included giving managers the right to fire workers. Institute academic councils gained greater responsibility for the financial success—or failure—of their respective institutes. The shake-out in scientific research, when combined with political and economic crisis, unfortunately led to high unemployment, shortages of hard currency, an inability to buy Western journals or equipment, and increasing brain drain to the West.[29]

The decreasing importance of ideology that accompanied economic and administrative changes accelerated the reform of scientific R&D during the Gorbachev years. First of all, primary party organizations in research organizations quickly lost power to institute academic councils. In the 1920s the *uchenyi sovet* (academic council) had great freedom to set research priorities and to comply with new reporting requirements imposed by the state. With the introduction of Stalinist science policy the *uchenyi sovet* was forced to pay

29. A. Golovkov and I. Lomakin, "Nauka: xhozraschë"t i gosudarstvennaia podderzhka," *Kommunist* 5 (March 1989): 86–89.

more attention to planning. Its work was overseen by a party official who had membership on the council. While still having relatively great latitude to deal with the day-to-day business of science, they lost much of their autonomy to central Party organs. They were also central in ensuring allegiance to the *partiinost'* (primacy of the party) and the *kollektivnost'* (collective). As Stephen Fortescue points out, since the 1950s, Party organizations have devoted less attention in their "agitprop" work to the imposition of Marxist ideology on the natural sciences, and Soviet attempts to distinguish "bourgeois" from "proletarian" science, which were particularly pronounced during the years of Stalin, all but disappeared.[30] As part of the de-ideologization and decentralization of science the academic council reasserted itself as the chief administrative body.

The Scientist and the State in the Former Soviet Union

In the aftermath of the failed August 1991 coup, the former Soviet empire unraveled completely. A commonwealth of independent states (CIS) arose in its place. Its future seems more shaky than ever. I will limit my comments to the impact of rapid political and economic changes on Russia and Ukraine since over two-thirds of the scientific institutions and personnel from the former USSR reside in those two nations. The general thrust of science policy in Russia and Ukraine today is based on the acceleration of the reforms initiated in the USSR until Gorbachev. Scientists and government alike realize that they are part of an international community. Many recognize that administrative apparatus for administration and funding of R&D must be democratized and decentralized. They believe that market mechanisms must play a role in determining the vitality and health of scientific institutions and research programs. The transition from the noncompetitive centrally planned system to a pluralist market-oriented one, however, has left Russian and Ukrainian science in dire straits. What is more, the same individuals who dominated scientific policy making in the Soviet Union have no intention of relinquishing power to younger, more progressive members of the scientific establishment.

In times reminiscent of the Russian Revolution of 1917, the Russian and Ukrainian scientific communities face a bleak future. Rampant inflation and shortages of hard currency preclude purchasing Western scientific journals, supplies, and airline tickets, or repairing machinery and equipment. Several major scientific journals have suspended publication. Many institutes have laid off or fired workers. Others have gone months without paying salaries.

30. Fortescue, *The Communist Party and Soviet Science,* (Baltimore: John Hopkins University Press, 1987). 40–41, 116–21.

All of this contributes to international isolation. Not surprisingly, brain drain is siphoning off top scientists in many fields. A vice president of the Ukrainian Academy of Science described the situation as "total collapse." Scientists speak without blinking of the need for billions of rubles and tens of millions of dollars.

Yet the reforms contributed to a new self-image for former Soviet scientists. Most important, scientists and officials from leading Academy and Ministry of Science personnel on down reject "go-it-alone" ideological pronouncements about the superiority of Soviet science. In post-putsch Russia, they welcome participation in international science through individual, institutional, and bilateral agreements. The Russian minister of science, Boris Georgievich Saltykov, a specialist in science policy, an economist by training, is such an internationalist. He calls for Russian scientists to seek out contacts to secure Western aid and employment. He does not fear "brain drain," but sees international flows of talent as part of the modern world. "It is better to work somewhere than nowhere at all," he said of Russia's scientists. "Let them go abroad."

The democratization of administrations has led to the proliferation of official and private organizations that play a role in science policy. In the past, prestigious Communist Party and scientific "bosses" made most decisions behind closed doors and dominated scientific life on the basis of personal connections. Higher authorities appointed institute and laboratory directors according to the *nomenklatura* (approved lists). Independent experts were seldom consulted. Now laboratory directors are elected, as are most institute directors with pro forma approval by the Academy presidium. Entrenched scientific bureaucrats were removed. As part of the fallout of the August coup, the Communist Party was outlawed.

Paradoxically, in the short term, the disappearance of the Communist Party from the scene has had a negative impact on fundamental research. The Party had played a central role in trouble-shooting in areas of equipment supply, access to current literature, travel arrangements, and so on. The usual discomforts of scientific research in the Soviet Union have become worse in the CIS without the Party as an expediter. Scientists are subject to the same gross inefficiencies that long handicapped other sectors of the Soviet economy.

Decentralization of power structures accompanied democratization. In the Soviet Union R&D was divided among the Academy of Sciences, where most fundamental research was performed; industrial ministries, where the bulk of applied research occurred; and the universities and other higher educational institutions with training functions. The State Committee for Science and Technology (GKNT), an agency whose power and influence exceeded its budget and staff size, attempted to coordinate the national R&D. In December 1991, GKNT merged with the Ministry of Higher Education and several other bureaucracies in the newly formed Ministry of Science, Higher Education, and Technology Policy of Russia, retaining 80 percent of former personnel.

The once powerful USSR Academy of Sciences, whose policies shaped the face of fundamental research for the entire empire, has given way to the Russian Academy of Sciences (RAN). The academies of the other states of the CIS have been left to determine their own paths. RAN was created in December 1991 on the basis of the Soviet Academy. All but one Soviet Academy institute and all but two of its 250 academicians came from Russian territory. The RAN added roughly 10 percent more members. The RAN is the most important scientific institution in the CIS by virtue of its historical legacy, and the number and excellence of its institutions and members.

Today, the democratization of the RAN lags somewhat behind that of society. The increased membership of the RAN is more nationalistic and conservative. The effort to give corresponding members and even doctors of science voting privileges has been remanded to a charter committee. In many cases, past scientific bosses still command great authority. But the RAN is no longer an institution of the state: by presidential decree the RAN has completely independent legal standing and property rights. The RAN provides its scientists and institutions with more equitable access to resources, including travel abroad. Rule by fiat, which characterized administration in the past, has come to an end.

One of the most striking changes in science in the CIS is the proliferation of groups of scientists independent of the government, what would be called "professionalization" in the West. In the last five years, independent scientific expertise has been reborn. We have seen the transition from Gorbachev's informal science advisory committee to the Yeltsin government's ministry of science. At least six new academies (including engineering, technology, humanities, economics, and agriculture), and dozens of scientific societies (from nuclear physics to sociology) have arisen in all states of the CIS. These societies—the Physics Society, the Nuclear Society, and the Union of Scholars, a kind of American Association for the Advancement of Science (AAAS), and many others—have three primary goals: to defend specialists' professional interests; to provide the government with independent expertise that would ensure well-informed decisions in science and technology policy: and to combat incipient antiscience attitudes among the population. However, most of the societies have fallen on hard economic times. What is more, the Nuclear Society and the "Chernobyl Union," a national, voluntary and independent social movement created in 1989 to prevent future atomic accidents, raise awareness of environmental issues, organize radiation education, and offer support to those who suffered in the Chernobyl disaster, have done little to dispel social concerns. This is because of the growing importance of the public in the political economy of science.

In the United States a loss of faith in scientists, engineers, and their work was associated with three events. The first was the stark contrast between the success of the Apollo mission and continuing poverty and poor race relations in the cities. Policymakers and many other citizens came to recognize the fact that even if you could put a man on the moon it did not necessarily mean you could similarly engineer all of society's social ills. Second, the Viet Nam War was shown on television, which graphically suggested that scientists produced weapons of horrifying technological accuracy but limited social utility. Third, growing awareness of environmental problems connected with pesticides and herbicides also soured public opinion toward science and technology. In the Soviet Union, loss of public support for technology was also connected with an expensive space program with few civilian spin-offs, war (in Afghanistan), and environmentalism. Certainly, glasnost and perestroika encouraged discussion of the appropriate relationship between technology and society, the former by revealing the extent of Soviet economic, political, and social problems, and the latter by encouraging examination of scientists, engineers, and their work and of how their research was approved, funded, and administered seemingly beyond social control.

Balancing scientists' newly found academic and political freedom are wide-ranging antiscientific attitudes, mistrust, and radiophobia among the public. From a nuclear power industry in critical health and spectacular space failures to daily press reports of newly discovered toxic waste dumps, the public has a growing awareness of the potential social costs of unregulated science and technology. The striking response is the proliferation of peasant home remedies, a fascination with medical charlatans who offer curative psychotherapy over the television, recently awakened latent interest in UFOs and extraterrestrials, and an ongoing urine-drinking craze. Such writers as Sergei Zalygin, editor of *Novyi mir,* Valentin Rasputin, and others representing the village prose genre have adopted a nearly Luddite attitude toward modern science. The public has embraced this view. Many citizens cannot understand how it is possible to fund big science—reactors, satellites, and particle accelerators—when consumer goods, food, and medical care are in short supply.

Scientist, and in particular physicists, have maintained a firm conviction, however, that the development of large-scale technologies must continue, indeed is inevitable, and sense the need to circle wagons to defend their work. Physicists continue to support extensive expenditures on nuclear, high energy, and space research. A skeptical, at times hostile public notwithstanding, physicists find it inconceivable to halt technological advance. A. D. Sakharov, known in his last years more for his political conscience than for his physics, pushed for the construction of nuclear power stations underground as a way around safety concerns and public resistance. But the Chernobyl disaster revealed the extent to which technological arrogance long held sway among nuclear engineers. Soviet engineers and planners pursued an aggressive program of commercialization of nuclear energy through the prefabrication of reactors and their components and such cost-cutting measures as less than adequate containment vessels. They believed that reactor technology is inherently safe, could be operated by nontechnical personnel, and that there is a technological fix for most problems.[31]

Conclusions

In this chapter, I have described the political economy of Soviet science. The polity changed from a system that closely approximated an authoritarian model to one of institutional pluralism to state reformism and the pluralist market orientation of today. From the founding of the Soviet state, the relationship between specialists, on the one hand, and party and government, on the other, was marked by tension reflecting attempts by the state to subordinate scientists' professional aspirations to its own programs, and by scientists' efforts to maintain a degree of autonomy in the face of this pressure. Under Stalin, the Party introduced a centrally planned, command economy, and a rigid, centralized system of governance. On one level, this led to domination of scientific research institutes by Party organs and an inefficient system for the determination of priorities and allocation of resources. On the next level, it contributed to the power of research institutes over the individual scientist, the sapping of the innovative impulse, and the domination of whole fields of research by one individual or institute. In addition, the state pressured the scientific community to emphasize applied research at the expense of fundamental research.

And yet, as this case study has tried to demonstrate, scientists, and physicists in particular, have been vocal actors in the political arena, contributing to Soviet economic and political development programs, and defending their discipline from ideological intrusions at each stage. They were able to find sufficient leeway within the Stalinist system to establish research priorities. Relatively generous budgets contributed to this, as did such social, cultural and ideological factors as the postwar cult of science that gave physicists some power and authority. Physicists assumed important social functions as visionaries of the Communist future in the Khrushchev years, and as the leaders of perestroika under Gorbachev. Thus, while dependent upon state support for their livelihood, they had skills central to the legitimacy of the state that included the development of the economy and the generation of

31. Paul R. Josephson, "The Historical Roots of the Chernobyl Crisis," *Soviet Union* 13, no. 3 (1986): 275–99.

international prestige. In short, as the Soviet Union became a modern state it needed increasingly to rely upon scientists for advice.

Many observers have emphasized the nature of the polity in the USSR as the major factor in determining the relationship between the Communist Party and scientists, at least until the contemporary period. They see the most important consideration as being Party domination of scientific research and development. As I have tried to demonstrate, this kind of analysis is too simple. It loses sight of the vitality of the relationship. Certainly, one of the most important influences on the development of postwar physics was the style of political leadership. Under Stalin, the Party was supreme. At each instance when physicists assumed that they had reclaimed control over some area of their discipline, the Party and its ideologues responded with further encroachment. Having reached an understanding about the relative importance of fundamental versus applied research on the eve of World War II, physicists found this equilibrium upset by the Nazi invasion and the wholesale disruption of their research programs, although they willingly and gladly contributed to the Soviet victory. Their hopes to parlay the recognition of their contribution into greater control over the physics enterprise, however, were roundly disappointed. In spite of increased material benefits—a reward for their war effort, they soon fell under attack for "idealism" and "cosmopolitanism" during the Zhdanovshchina.

On the eve of Stalin's death, nonetheless, Soviet physicists stood poised to take advantage of the support of the Communist party for rebuilding an economy devastated by war to acquire significant increases in funding for fundamental and applied research. By then, Soviet physics occupied a leading position in the international scientific community in nuclear, solid state, and theoretical physics. Largely on their own initiative, physicists had organized on a national scale, founded a series of relatively well-equipped research institutes, and were often able to balance the Party's demands for more research of an applied nature with their own programs to strengthen fundamental science.

Under Khrushchev, the relationship between physicists and political power changed for all time. Khrushchev recognized the importance of advances in science and technology as crucial factors in the construction of "the material basis of Communism." He took advantage of Soviet successes in big physics to bolster his own position within the Party. He supported independent scientific research within the constraints of the Soviet system. For their part, physicists utilized Khrushchev's support to begin to reestablish control over their discipline. They played an increasingly active role in the determination of national science policy, which culminated in newly found political power under Gorbachev. The reforms initiated under Gorbachev enabled scientists to remove entrenched bureaucrats and displace Party organizations as the seat of scientific policy-making.

A major factor contributing to the resurgence of scientists' political power has been the decline in the importance of ideology. Having suffered from ideological interference since the mid-1930s, physicists wished to demonstrate once and for all that theoretical physics was compatible with dialectical materialism. By participating in the development of a "cult of science" and "atomic culture," physicists soon sold the Party and populace alike on the notion that scientific and technological achievements were the key to the future, and that they would demonstrate the superiority of the Soviet system. The successes in nuclear weapons, nuclear power, larger and larger particle accelerators, and the space program in particular confirmed this fact. During the de-Stalinization thaw they reclaimed control of this area of their discipline. During the two decades of Brezhnev's rule, a conservative ideology dominated Soviet society, and, although it did not have a significant impact on the philosophy of science, it was sufficient to derail the reformist tendencies of the Khrushchev years. Under Gorbachev, the emphasis on individual initiative and market mechanisms in science has put an end to the centrality of *partiinost'* and *kollektivnost'* in science.

The most striking feature of the political economy of Soviet physics was a characteristic it shares with other countries. Since the founding of the Soviet state, Party leaders, ideologues and physicists alike have believed that technology is the key to economic development and Soviet success. They have argued that technological innovation is always desirable, and that technology is infallible, a symbol of national achievement, and the highest level of culture. Large-scale, expensive technologies that required highly centralized and extensive networks of support were seen as safe and desirable, as solutions for social and economic problems. This factor more than any other may have contributed to the ability of physicists and the party to reach agreement on research priorities and national science policy.

Scientists in Russia and Ukraine today believe that the end of the Soviet empire means the end of ideology in science. They see scientific activity as an apolitical endeavor that seeks to identify value-neutral and objective laws of nature. They see the Western system of science, particularly that in the United States, with its relatively decentralized funding, administration, and market mechanisms, as a panacea for the crisis of funding and decay they now face. It may be desirable for them to read the chapters in this volume, and to contemplate the ways in which scientific activity is influenced by the political economy of every system.

Brazil: Scientists and the State—Evolving Models and the "Great Leap Forward"

Simon Schwartzman

In the 1970s, the Brazilian military government decided to make the country join the exclusive club of world powers and, for a brief period, placed science and technology at a level of priority it never received before, nor since, in the country's history. Yet, the ability governments have to obtain results in science and technology depends on previous developments they do not control. In what follows, we shall examine in some detail the assets Brazil had for its ambitious attempt, why the attempt ultimately failed, and the consequences it left in terms of Brazil's current scientific and technological ability to face the realities of the 1990s.

Historically, Brazil has moved back and forth between stronger and weaker governments, democratic and authoritarian regimes, liberal and nationalist policies. This wavering makes it difficult to classify the country as either market-oriented or centrally planned, politically competitive or authoritarian, in terms of Etel Solingen's typology in the introduction of this volume. In broad terms, however, there are reasons to point to a strong tendency toward political and administrative centralization and a political split between the political and administrative elites (including the military), based on the administrative centers, and the more entrepreneurial elites based in São Paulo and other southern regions (Schwartzman 1988a). The political split between the administrative and economic elites has also led to the differentiation between two sectors in the economy: one based on private capital, often in association with foreign firms; and another based on a few large state-owned companies (oil, electricity, telecommunications, ore extraction, steel production, energy, railroads, and so forth). Besides, the Brazilian government controls a large sector of the country's financial system, through the Banco do Brasil, Caixa Econômica Federal, Banco Nacional de Desenvolvimento Científico e Tecnológico, and a network of state banks. The weight of these public companies have made the Brazilian public sector the dominant one in the economy, forcing the private sector to depend on public contracts, facilities, and privileges to survive. These public corporations were crucial for the country's extremely high rates of economic development up to the late 1970s, with average growth rates of about 7 percent a year. Yet, in the 1980s, they became burdened by increasing overhead costs, heavy indebtedness and low productivity, and the extinction, privatization, and streamlining of public corporations has been an important goal of Brazil's new governments after 1990. In that sense, it is possible to say that Brazil is moving from a non-pluralist, semimarket economy, only partially open to competition, to a pluralist market-oriented society with a smaller state and a stronger private sector. This is the framework in which the developments described subsequently have occurred, and are still being played.

Preconditions: From Benign Neglect to Ritual Confrontations

Brazilian science and technology developed late, and its aspirations and predicaments seldom reached the country's higher levels of government—it is a long history of benign neglect, with a few bright spots of government awareness and scientific and technological achievements. One of these spots was the second half of the nineteenth century, when Emperor D. Pedro II, an enthusiast of modern inventions, provided for the maintenance of a few natural history museums, the astronomical observatory, botanic gardens, and attended personally to academic events in Rio de Janeiro's schools of medicine and engineering. At the time, the country's economy was mostly centered on coffee and other plantations based on slave labor, and no thought was given to the notion that science and technology could in any way contribute to modernizing it.[1] Eventually, slave labor gave way to the work of immigrants, and the old Empire was replaced by a decentralized Republican regime with power in the hands of the rural oligarchies of the country's two main states, Minas Gerais and, above all, São Paulo.

Nineteenth-century imperial science was replaced by a period dominated by applied work in agriculture and health. Endowed with large resources provided by an expanding international coffee market, the São Paulo elites created a host of teaching and research institutes at the turn of the century, later populated by the children of immigrants from Europe, Japan, and other Brazilian states who flocked to the region. Agricultural expansion brought about a few centers for applied research—the Instituto Agronômico de Cam-

1. This is not completely true, since the old "botanical gardens" were meant as acclimatization stations for plants imported from India. Mining, however, received more attention, with less significant results. The most important initiative in this regard was the creation of the Escola de Minas de Ouro Preto in the 1870s (Carvalho 1978).

pinas (agronomy), the Instituto Biológico, the Instituto Butantã, famous for its work on snake venom, the Instituto Bacteriológico (Bacteriological Institute,) and the state's Comissão Geográficà e Geológicà (Geographical and Geological Commission). The histories of these institutions are tales of important achievements, but also of the difficulties they had in establishing long-term, quality scientific research, their isolation from the professional schools, their reliance on foreign leadership,[2] and the difficulties these foreigners had in carrying out their work. The most significant achievement in the period was not in São Paulo, but in the country's capital, Rio de Janeiro: it was the consolidation of the Oswaldo Cruz Institute as a first-rate center for bacteriological research and tropical disease control.

The next period, starting in the 1930s, can be characterized by the search for new alternatives, and the beginnings of university-linked research. In spite of the Oswaldo Cruz Institute's achievements, Rio de Janeiro remained too narrow a basis for more comprehensive developments, and in the 1930s São Paulo would establish its permanent place as the country's most significant center for higher education and research. The achievements in São Paulo can be explained by a combination of two factors: first, the basis provided by a dynamic economy centered on cash-producing crops (coffee above all), an emerging industrial park, hundreds of thousands of European and Japanese immigrants, several professional schools (some of them of reasonable quality), and its centers for applied research; second, the initiative by the São Paulo elite to create, in 1934, a new university based on the School of Sciences (the Faculty of Philosophy, Sciences, and Letters) staffed with foreign scholars, which inaugurated the practice of academic, university-based research in Brazil (Limogi 1989; Schwartzman 1991, chap. 5).

Research in high-energy physics was probably the most visible product of the new institution. A couple of Italian professors in physics and mathematics (Gleb Wataghin, Giuseppe Ochiallini, and Luigi Fantapié) with adequate knowledge of what was happening in Europe, and the active recruitment of a few bright young people attracted by the possibilities of fellowships for studying abroad, were enough to introduce modern physics to the country, with surprising consequences for the next decades.

The postwar period, up to the late 1950s, is marked by an effort by the scientific community to assert its role in Brazilian society, while receiving from government little more than benign neglect. In those years, the Brazilian Society for the Advancement of Science was organized in São Paulo, the

---

2. Among others: F. W. Dafert at the Estação Agronômica; Hermann von Ihrering at the Museu Paulista; Orville A. Derby, Albert Löfgren, and F. C. J. Schneider at the Geographical and Geological Commission; Adolph Lutz at the Bacteriological Institute. For more information on Dafert, see Dean 1989.

first of a growing number of specialized scientific associations (Botelho 1990; Fernandes 1987); and, in the early 1950s, the Brazilian government put in place the National Research Council, which was meant, at first, to provide support to an emerging atomic research capability (Albagli, 1987; Cagnin and Silva 1987; Romani 1982).

After World War II atomic research was on everyone's mind, and Cesare Lattes, one among the Universidade de São Paulo's young physicists, had gained notoriety by participating in Ochiallini's work leading to the empirical determination of the meson pi in 1947. The São Paulo's physicists had cooperated with the military in a few technological projects during the war, the most important being the development of a sonar for submarine detection. It is not difficult to imagine how the idea of bringing these experiences together in a big atomic energy project came about. The project required an important military name, which was provided by Admiral Álvaro Alberto de Mota e Silva; a new institutional setting, this time in the country's capital; and a highly prestigious scientific figurehead, Cesare Lattes. In 1949 the Brazilian Center for Physics Research was established in Rio de Janeiro, followed two years later by the National Research Council, which included an embryonic National Commission for Atomic Research. The whole project did not go very far. The United States government did not look favorably to the project and intervened actively when the Brazilian government tried to buy some ultracentrifuges for uranium enrichment from Germany.[3] Brazil had physicists who understood atomic theory, but knew very little of atomic energy technology. Whatever governmental support the project had, it disappeared when Brazil's president Getúlio Vargas committed suicide under pressure in 1954.[4]

After that, the National Research Council lingered as a small outfit providing money for individual research projects in the natural sciences, and, with a few exceptions, nothing much happened for the next ten years or so.[5]

---

3. In the seventies, during the Ernesto Geisel government, the United States would oppose again a deal between the Brazilian and the German governments in the field of atomic energy.

4. A less-known project for nuclear fuel self-sufficiency, which was developed by the so-called thorium group at the Federal University in Minas Gerais, came to a halt at the same time. Participants in this group later became involved in the formulation of the alcohol program, which aimed at replacing petrol as the main source of fuel for internal combustion engines.

5. An important exception, however, was the establishment of a school of aeronautics engineering, the Instituto Tecnológico da Aeronáutica (ITA), in the São José dos Campos region between Rio de Janeiro and São Paulo. ITA was an exceptional institution on many accounts, from its freedom from the Ministry of Education's bureaucracy to its strong links with North American institutions, above all MIT (it even had a U.S. rector for several years). Although a military institution, its students were mostly civilian, and they were recruited nationwide through stringent competition. ITA became famous for its teaching standards, many of its students developed brilliant careers as engineers, scientists, and managers. It was also at the origin of Brazil's aeronautics industry; but, curiously enough, ITA was never renowned for its scientific or technological research.

Unable to exert influence at the top, many scientists concluded that only political mobilization from the bottom could bring them to the important roles they considered their responsibility to play. Many became enrolled in political parties on the left, and joined the student movement in campaigns for university reform. The early 1960s were a time of intense political, and not always ritual, confrontation. In 1964 the military took control of the government, and started a long and intermittent process of political purges that would come to a climax in 1969, when hundreds of scientists and university professors, some of them of international prestige, lost their posts and eventually left the country.

The "Great Leap Forward"

While, in the late 1960s, conservative officers at the intelligence agencies were busy hunting for Communists and suspicious intellectuals, others in planning and education agencies were thinking on how not to waste the country's limited scientific talent and how to build new and stronger scientific and academic institutions. The involvement of Brazil's main investment bank, the government-owned Banco Nacional de Desenvolvimento Econômico, in the field of science and technology after the mid-1960s, is the most important feature of the new period.[6] For the first time in Brazil's history, there was a concerted attempt to put science and technology at the service of economic development through the application of a substantial amount of resources. In 1964 the Banco Nacional established a program for technological development known as the Fundo Nacional de Tecnologia (National Fund for Technology), which in its first ten years provided a total of about 100 million dollars for research and graduate training in engineering, the natural sciences, and related fields.[7]

The Fundo Nacional started with the hope that economic incentives could lead private investors to develop their technologies, instead of importing them from abroad; very soon it began to support selected teaching and research programs. This change is partially explained by the absence of coordination or

6. The following section is based on Schwartzman 1991, chap. 9.

7. Another development, not covered here, was the establishment of research departments and units within the largest state-owned corporations, such as Petrobrás, Telebrás, and Companhia Vale do Rio Doce. No proper evaluation of these activities was ever done, and, working within state-controlled public corporations, these research units were protected both from the academic competition typical of universities and from cost considerations derived from market pressures. The little evidence available suggests that these research institutions, although better endowed with equipment and salaries than many others, did not perform as well (Schwartzman 1986). There was also a large and still not fully evaluated development in the field of agricultural research that has been run by an independent agency, EMBRAPA, under the Ministry of Agriculture.

shared goals between the economists belonging to the Ministerio de Planejamento (Ministry of Planning), who pushed for technological development and self-reliance, and those at the Ministerio de Fazenda (Ministry of Treasury), which actually decided about the country's economic policies.[8] With the Fundo Nacional's support, the Universidade de São Paulo acquired its electrostatic accelerator, Pelletron, in 1971; a consortium of institutions started the development of a Brazilian minicomputer; the Centro Tecnológico da Aeronáutica obtained support for its development of airplane engines; the Instituto Militar de Engenharia was able to start its graduate programs in several branches of engineering and chemistry; and the newly created Universidade de Campinas received substantial grants for a variety of projects, mostly in the field of solid state physics. An important initiative was the creation of a complex system of graduate courses in engineering at the Universidade Federal do Rio de Janeiro, known by the acronym COPPE. The activities of the Fundo Nacional were later transferred to a new, specialized agency, the Financiadora de Estudos e Projetos (FINEP), working as an investment bank for technological and feasibility studies and administering a national fund for science and technology that became part of the federal budget in replacement of the development bank's fund. While some of the best-known scientists expelled in the previous years remained critical of the whole undertaking, and were often unable to participate, most of the young generation and the less conspicuous names joined in, without having to endorse the conservative views of the military in matters of social and political rights.

The entrance of agencies of economic development and planning into the field of science and graduate education intensified the historical tendency to favor applied technology over basic science. In 1975 the old Conselho Nacional de Pesquisas was transformed into a new and much larger Conselho Nacional de Desenvolvimento Científico e Tecnológico (National Council For Scientific and Technological Development, CNPq), now under the Ministry of Planning. Science supporting agencies like FINEP and the CNPq evolved gradually into swollen bureaucracies of hundreds and eventually thousands of functionaires, where scientists negotiated with economists and administrators every two or three years, on a project-by-project basis, for the renewal of their grants. A two-year national plan for science and technology was promulgated in 1973, and again in 1975, projecting expenditures ranging from 323 to 824 million dollars a year (Schwartzman 1978). These plans were little more than aggregations of expected expenses by sector, most of which—65 percent for the 1973–75 period—were completely outside the sphere of influence of the

8. The Ministry of Industry and Trade, nominally in charge of industrial development, was never an important actor in economic, technological, or industrial policy decisions, but it had a limited influence on the beginnings of the alcohol fuel program. (Schwartzman and Castro 1984).

planning authorities, in agencies such as FINEP, the National Research Council, or the Banco Nacional de Desenvolvimento. The expectation for the 1976–77 period was to increase slightly the amount under these agencies. Between 21 percent and 27 percent of the expenditures were to be allocated to graduate training, fellowships, and "scientific development" in general; between 20 percent and 29 percent to industrial technology; between 11 percent and 15 percent to agricultural research; and between 5 percent and 10 percent to atomic energy projects. There is no known evaluation of how the plan was implemented, nor of how the expenses were made.[9]

The new emphasis on research and technology coincided with a reform of the country's higher education system that introduced several features of the U.S. system, including full-time teaching and research, and graduate education. The strategy adopted by the science and technology agencies was to identify what they considered good or promising research groups in universities and to provide them directly with support, very often by-passing the established procedures for labor contracts, accounting procedures, and decision-making within the universities. For the researchers, there was now a "market" open for projects and proposals that was sensitive to their qualifications and aspirations. For the universities, it meant that new resources became available, but also that they flowed completely out of their control. Well-equipped, well-staffed, and well-paid departments and research programs started to exist side-by-side with poor ones, the former more concerned with research and graduate education, the latter bound to the traditional undergraduate schools and courses.

From the planning agencies' perspective, the strategy worked well. In 1970 there were fifty-seven doctoral programs available in Brazilian universities; in 1985 there were more than three hundred, with another eight hundred providing training at the M.A. level (Paulinyi et al. 1986). About 90 percent of these courses were in public universities, graduating about five thousand students at both levels each year. On all accounts, Brazil started to build a fairly significant scientific community.

The culmination of this drive was the establishment of brand new institutions that would be free from the limitations of the past. These institutions ideally should be as free as possible from institutional and bureaucratic limitations or restrictions; they should be able to receive large amounts of money from science planning agencies and put the money to work in the hands of well-qualified people; and they should work in the frontier of the modern technologies that the country supposedly needed for its economic and indus-

9. The third plan, for 1978–79 under President João Batista de Figueiredo, came out in a period of economic crisis and political unification of the planning and treasury ministries under Delfim Netto and was just a broad statement of purposes without any figures attached.

trial development. Two institutions, more than any others, met these requisites: the Universidade de Campinas and the engineering program of the Universidade Federal do Rio de Janeiro, COPPE. They had in common a strong, individual leadership (Alberto Luís Coimbra and Zeferino Vaz), freedom from routine bureaucratic procedures, and direct support from the science and technology development agencies for their ambitious projects.

A full analysis of this experience of forceful scientific and technological development would have to include the whole field of agricultural research, some achievements in biotechnology, the development of the airplane and arms industry, the technology for steel production (Dahlman and Fonseca 1987), and the role of institutions such as the Instituto de Pesquisas Tecnológicas in São Paulo. However, two cases of high technology, atomic energy and computers, stand out for their proximity with the basic sciences as well as for many other reasons, and require a closer look (Adler 1987; Schwartzman 1988b).

We have seen how in the mid-1950s, confronted with the alternatives of trying to develop its own technology with the help of the existing scientific community or the acquisition of foreign technology, the Brazilian government chose the second. In 1975 an ambitious agreement for nuclear cooperation was signed with West Germany that included the construction of several nuclear energy plants and the transfer of enriched uranium technology. The agreement drew strong opposition from Brazilian scientists because it consisted mainly of the transfer of engineering technology and did not incorporate the acquired or presumed competence of Brazilian scientists. With the oil and debt crisis starting in the late 1970s, the agreement proved to be overambitious, and it is now limited to the construction of two power plants at most, neither of which is near completion as of this writing.[10]

A so-called parallel program of atomic research was also undertaken by the Brazilian military, outside the restrictions built into the German-Brazilian agreement. Rumors that Brazil is developing its own atomic bomb have never been confirmed; but the government has acknowledged the development of nuclear engines for ships and submarines, and in September 1987 it was formally announced that Brazil had developed all the needed technology for the production of nuclear fuels for pacific purposes. The method of ultracentrifugation was said to be similar to the one used by the URENCO consortium in Europe, and the grade of enrichment, which was announced to be of 1.2 percent, was supposed to increase to 20 percent in one or two years, when an industrial plant was also supposed to start working. The work had been

10. There is only one working atomic energy plant in Brazil, built by Westinghouse in Angra dos Reis before the German agreement, known for its successive technical difficulties and overrun costs. Large investments have been made on a sister plant that has remained inconclusive.

carried on for eight years at the Instituto de Pesquisas Nucleares from Universidade de São Paulo with support provided by the Brazilian Navy, at a stated cost of 37 million dollars. The announcement was received with widespread skepticism by the Brazilian press and Brazilian scientists. The level of enrichment was considered too low for any practical purposes; the expenses too high, given the country's deep economic crisis and the other needs of scientific institutions; and the concentration of resources in military research, at the expense of civilian institutions, was seen as a worrisome trend.

While a key feature of the nuclear program was the exclusion of university-based scientists and the development of large, state-controlled bureaucracies, the computer industry started from people coming out of the universities, and gave rise to several privately owned companies that grew under the umbrella of market reserve (Adler 1987; Frischtak 1986; Schwartzman 1988c). This policy coincided with the worldwide explosion of the microcomputer industry. By the end of 1984, at the closing of the military regime, a bill defining a "national policy for informatics" was approved by the Brazilian Congress by a majority of votes from the government and the opposition parties.

There were many remarkable elements in this policy, which generated many studies, international press coverage, and threats of retaliation from the Reagan government. The main novelty was that this was the first attempt in Brazil's history to develop an industrial policy based on local technology and purely Brazilian firms. The Brazilian policy for computers is an extreme case of "backward integration," in which production starts with the assembly of the final product with imported parts, with the expectation that the product would incorporate an increasing amount of locally produced components (Nau 1986). Such a policy would require, among other things, a corresponding investment in research and development (R&D), which in the Brazilian case did not exist. The weakness of the research effort is one reason why the Brazilian policy for the computer industry came under severe criticism not only from foreign competitors hoping to sell in the Brazilian markets, but also from end-users and manufacturers in the country who see this policy as a growing roadblock to their access to state-of-the-art technologies. The market protection policy for microcomputers was officially abandoned by the Brazilian government in 1992. In computers, as in atomic energy and in other applied fields, the great leap forward was far shorter than the original expectations.

Downhill: Realities and Expectations for a Ministry of Science and Technology

Brazil was unfortunate in that political freedom and democratic rule started to be implemented precisely when economic crisis stuck, in the

1980s.[11] The Ministry of Science and Technology was established by the new civilian regime in 1985, in response to demands from a significant section of Brazil's scientific community. The Brazilian Society for the Advancement of Science, SBPC, produced at the time a document outlining the critical issues in the country's scientific institutions, and the hopes for the new era. The expectation was that the new agency could at least maintain and strengthen the country's scientific and technological research institutions, so that they could not be destroyed or maimed, as it happened so often in the past, by indiscriminate measures of budgetary restraint, inconsiderate bureaucratic decisions, or sheer political patronage. This role of protecting the country's scientific assets was to be accompanied by new institutional mechanisms to strengthen the research institutions, stimulate academic and scientific exchange and cooperation, and turn basic research into an effective instrument of social and economic development. The Ministry of Science and Technology was supposed to articulate a long-range policy for scientific and technological development, with the scientific community's help, for the approval of a freely elected Congress. Finally, there was the expectation that the new ministry could help in the definition of a "new and well-defined industrial policy, with economic incentives to companies willing to develop national technologies and financing instruments adjusted to the risks involved," as a way of overcoming the country's economic and technological dependency from outside (Comissão das Sociedades Científicas, 1985).

The common understanding was that, to carry out these tasks, the new ministry would need to be closely watched and helped by the country's scientific community. In fact, interactions between scientists and government underwent significant transformations since 1985. Peer review has been a longtime tradition at the Council and at the Ministry of Education. Before 1985, however, participation of scientists in collegiate bodies was always by invitation, or cooptation. After 1985 these two agencies developed elaborate systems for the nomination of advisers and reviewers by scientific associations and research groups, and the very presidency of the National Research Council was given to a former president of the Brazilian Association for the Advancement of Science. A new interministerial body for science and technology coordination, the Council of Science and Technology (CCT) was also instituted, with the presence of government ministers and representatives of the scientific community.

This trend of increasing representation of scientists on decision-making bodies came under difficulties and criticism on different and sometimes con-

11. It was probably not just a coincidence; the gradual steps taken by the Figueiredo regime (1978–85) to return the country to civilian rule have been interpreted as a kind of deliberate decision by the military for strategic withdrawal in face of the failure of the grandiose development plans of the previous years.

flicting grounds. The swollen bureaucracies in the agencies resented the scientists' interference with their decision powers, and tried to develop alternative mechanisms for their action; well-established scientists questioned the impartiality of reviewers selected through institutional and regional affiliation, rather than strict academic criteria; new, emerging, and less-qualified groups were afraid that reviewers from the more developed regions and institutions would concentrate resources in their own hands; economists and civil servants complained about the dispersal of resources and called for more planning and the establishment of priorities; while traditional politicians pressed for personal and partisan criteria in the distribution of benefits such as fellowships and research grants.

The net result was that, in the new and expanded political arena of the postmilitary regime, the scientists had to fight for access and control of science policies, agencies, and resources with much more and unexpected partners than they imagined. Budgetary data for 1987 show that the Ministry of Science and Technology controlled only about 30 percent of federal expenditures defined as R&D; another 20 percent belonged to the Ministry of Agriculture; 14 percent went to the Ministry of Education (mostly for graduate education); and 16.5 percent remained directly under the president. Seen from another perspective, 60 percent of the resources in that year were linked to about twenty large projects or programs. Of this 60 percent, 36.6 percent were attributed to military projects, 18.7 percent to agricultural research, and less than 30 percent to the National Research Council and the National Fund for Scientific and Technological Development (table 1). No coordination existed among these different ministries and projects, and the interministerial council responsible for establishing a global policy and budget never acted in that capacity. Besides the scattering of R&D resources throughout the federal government, science policy decisions within the realm of the new Ministry became subject to a variety of other influences beside that of the scientists—bureaucrats, regional interests, teachers' and employees' associations and unions, politicians, lobbyists for technologically intensive industries, and so forth. As soon as it was established, the new Ministry started to build a complex bureaucracy for science and technology planning in all areas of the so-called new technologies, with no provision for regular peer review evaluations except for occasional invitation.

Four years later, at the end of José Sarney's presidential period, few of the expectations surrounding the new ministry of science and technology had been fulfilled, and the problems affecting the country's scientific and technological institutions were much worse than before.[12] It is true that political restrictions to scientists and their work had ceased to exist. There was also, at the beginning, a

12. The following is based on Schwartzman and Reis 1989.

TABLE 1. Brazil, Projected Expenditures for Science and Technology in the Federal Budget, 1987

Project	Value in Cz$ 1.000	Percentage among the Large Projects	Percentage of Total S&T Projects
Advanced research projects related to National Security, National Security Council	1.000.000	10.1	5.8
Space and satellite development program, Armed Forces High Command	1.283.800	13.0	7.5
Applied and basic research, National Commission of Atomic Energy	398.600	4.0	2.3
Aerospace R&D, Ministry of Aeronautics	476.500	4.8	2.8
Sectoral Program on Ocean Resources, Ministry of Navy	115.000	1.1	0.7
Antarctic Mission, Ministry of Navy	260.000	2.6	1.5
Submarine project, Ministry of Navy	103.800	1.0	0.6
Applied research in Agriculture, EMBRAPA, Ministry of Agriculture	1.637.125	16.5	9.6
Graduate education fellowships program, CAPES, Ministry of Education	576.882	5.8	3.4
Development of agricultural technology for the improvement of sugar cane crops, Ministry of Agriculture	126.203	1.2	0.7
Research on natural resources, Superintendence for the Development of the Amazon Region, Ministry of Interior.	100.000	1.0	0.6
Research and experimental development, Oswaldo Cruz Institute, Ministry of Health	245.262	2.5	1.4
Studies and research for transports planning, Brazilian Company for Transports Planning, Ministry of Transports	114.600	1.1	0.7
Basic research, National Research Council, Ministry of Science and Technology	333.340	3.4	1.9

(*continued*)

Table 1—*Continued*

Project	Value in Cz$ 1.000	Percentage among the Large Projects	Percentage of Total S&T Projects
Applied research, National Research Council, Ministry of Science and Technology	200.000	2.0	1.2
Current expenditures of seven research institutes and national laboratories of the National Research Council	335.599	3.4	2.0
Graduate education fellowships, National Research Council	442.080	4.5	2.6
Research fellowships, National Research Council	382.000	3.9	2.2
Program of Energy Mobilization, general expenditures of the federal government	663.905	6.7	3.9
National Fund for Scientific and Technological Development, Ministry of Science and Technology (FINEP)	1.114.122	11.2	6.5
Total	9.909.033	100	58.8
Other expenditures	7.176.967		41.2
Grand total	17.086.000		100

Source: Ciência e Tecnologia, Orçamento da União 1987, MCT, CNPq. (The figure for FINEP, part of the overall budget of the Ministry of Science and Technology, was obtained directly by FINEP). Conselho Nacional de Desenvolvimento Científico e Tecnológico, Ministério da Ciência e Tecnologia, *Orçamento da União,* 1987. Brasilia, CNPq/MCT, 1988.

substantial expansion of resources for the National Research Council (mostly through an increase in the number of fellowships available for researchers and graduate students), significant increases in the salary levels of university professors and researchers in the federal system, and an ambitious program of fellowships for graduate education abroad was announced.

Seen from hindsight, it is clear that the resource expansion in those years did not follow a well-defined governmental commitment with science and technology, but was just another item in a general pattern of uncoordinated expenditure expansion, one cause of uncontrolled inflation in the succeeding years. The rapid expansion of the number of fellowships for studies abroad, to give an example, was not accompanied by any measure to create new working opportunities for returning graduates, generating a potential for increasing brain drain. In another example, although the government kept the letter of the legislation defining a national policy for the computer industry based on market protection and scientific development in the field, no industrial policy was devised to follow this legislation, nor was any priority given to R&D activities in the field.

The Ministry of Science and Technology was never perceived by the federal government as an agency to carry on specific policies, but just as a concession to be given to a specific sector of the PMDB party that provided it with political support. For the government, scientists and technologists were just another lobby pressing for their private interests, to be contemplated according to their political clout. For a while, within its narrow limits, the new Ministry and its holders could do more or less as they wished and benefit simultaneously from the expansion of public expenditures. They lacked influence, however, on several of the most significant sectors of technological research, such as agriculture, nuclear, space, and on industrial and educational policies. As the economic crisis increased, and the political alliances supporting José Sarney eroded, the Ministry could not protect the scientific institutions from budget slicing, barriers to the acquisition of instruments and materials abroad, and restrictions to international travel for scientists. More seriously, it became obvious that the new Ministry was not immune to the practices of pork-barrel politics prevailing in other sectors of the federal administration.

The picture became clear when, at some point in 1988, the political alliance between the Sarney government and the more militant sector of the PMDB party broke down, leading to the nomination of a new minister of science and technology with little or no links with the more active sectors of the scientific community. One of the first acts of the new minister was to issue a decree withdrawing all decision powers from the academic council (Conselho Deliberativo) of the National Research Council, which led to an intense mobilization of scientists and a retreat from the government on that measure. From that point on, the behavior of the Sarney government regarding the area became extraordinarily erratic. First, the minister agreed to go back in his decree and to open a discussion with the scientific community on the reorganization of the Research Council; then the Ministry itself disappeared, absorbed by the old Ministry of Industry and Trade, to be later reestablished as a "special secretary" for science and technology under the president; and finally, in December 1989, it was reinstated again as a full-fledged ministry. The administrative reform introduced in early 1990 by President Fernando Collor reinstated the Secretary of Science and Technology roughly with the same attributions, and never granted it a significant role.

Possible Trends

The crisis that affects the scientific community in Brazil in the early 1990s is likely to be deeper than the simple consequences of an incompetent government and runaway inflation. There are signs suggesting that the pattern of scientific development started in the late 1960s (and agonizing since the early 1980s) is approaching its end, without any clear indication of what will replace it, or even whether it will be replaced at all.

A first sign in this direction was the drastic reduction of the federal budget for science and technology since 1989, coming almost to zero (except for fellowships and money from a World Bank loan) in 1991–1992. There was some expectation that this withdrawal of the federal government from science and technology could be partially compensated by the creation of several budgetary funds and foundations for science and technology in many Brazilian states. In the past, only São Paulo had its science foundation, the Fundação de Amparo à Pesquisa de São Paulo, whose budget increased from 1/2 to 1 percent of state-level taxes in 1990.[13] State constitutions approved throughout the country in 1989 provide fixed allocations going up to 3 percent of state budgets or taxes in many other states. In some places, like Minas Gerais, Rio de Janeiro, and Rio Grande do Sul, science foundations similar to FAPESP were organized; in others, the money remained at the discretion of the state governor or some political appointee. Few states will be able to reproduce the institutional and ethical conditions that made FAPESP a highly effective and praised institution, strong enough to withstand the opposition or indifference of a populist governor. Even in the best scenario, the new money being provided for research at the state level will not compensate for the shrinkage of federal resources, and interregional inequalities in R&D allocations will tend to increase (this pattern of reduction of federal expenditures without compensatory allocations at the state and local levels is a general feature of the Brazilian 1988 Constitution, rather than something specific of the science and technology field).

A second sign is the loss of legitimacy of academic research and higher education as worthy recipients of federal funds, in the name of applied R&D and basic education. The designation of José Goldemberg, the former rector of the Universidade de São Paulo and a well-known physicist, to the Secretary of Science and Technology (and later Ministry of Education), by the Fernando Collor government, was a gesture signaling the new government's commitment with the strengthening and improvement of the science and technology sector. In practice, the money available for scientific and technological research continued to shrink, and successive statements about the need to give more emphasis on applied work were not followed by specific policies. From the beginning, short-term economic and administrative issues, followed by crisis of legitimacy, monopolized the new government's attention, and there

13. FAPESP's budget for 1991 was about 114 million U.S. dollars of which 82 percent is direct support for research projects, and 9 percent is for fellowships.

are no definitions about the mechanisms, speed, and investments that could shape new policies for the years to come.[14]

The Brazilian economy is going through a deep readjustment, likely to last for several years, that will keep public expenditures down, and open the economy still further for foreign investments and capital. This trend does not bode well for industrial policies based on market protection and technological self-sufficiency, as existed for the computer industry; and there are no signs that the country is ready to develop technology-intensive industries able to survive in a world of increasing competition, with or without government incentives. Public higher education, which has been stagnant since the early 1980s, is likely to remain so in the future. One consequence has been the reduction of the labor market for graduate degree holders, with negative implications, in terms of demoralization and lack of stimulus, for the research graduate programs at the universities.

If this scenario is correct, the scientific community is likely to react through increasing mobilization and confrontation against the federal government. The ultimate fate of this confrontation will depend on whether the government will succeed in its economic adjustment measures. If it fails, the opposition it will get from the scientific community will not be different from the opposition it will get from a wide range of organized sectors in Brazilian society from trade unions to public employees, and from teachers' associations to the press. Even if it succeeds, nothing will be like before. Brazilian scientists will have to live in a much tighter reality, where their projects will have to be justified for their economic, social, and educational worth, in face of stiff international competition. This will be a wholly new experience, for which neither the Brazilian government nor the scientists are prepared to face.

REFERENCES

Adler, Emanuel. 1987. *The Power of Ideology: The Quest for Technological Autonomy in Argentina and Brazil.* Berkeley: University of California Press.

Albagli, S. 1987. "Marcos Institucionais do Conselho Nacional de Pesquisas," *Perspicillum* (Rio de Janeiro, Museu de Astronomia e Ciências Afins), 1 (May): 1–166.

Botelho, Antonio Jose Junqueira. 1990. "The Professionalization of Brazilian Scientists, and Brazilian Society for the Progress of Science (SBPC), and the State, 1948–60." *Social Studies of Science* 20:473–502.

14. In March 1992 the new Secretary for Science and Technology, sociologist Hélio Jaguaribe, announced that the government would give two-thirds of its research money to applied work, and one-third to basic sciences. History shows that this preference for applied work has been the pattern all along; the problem is the ability of the productive sector to absorb and benefit from these incentives, if they materialize.

Cagnin, M. A. H., & D. H. Silva. [1967] 1987. *A Ação de Fomento na História do CNPq*. Brasília, Ministério da Ciência e Tecnologia. Reprinted Brasília: CNPq.

Carvalho, José Murilo. 1978. *A Escola de Minas de Ouro Preto: O Peso da Glória*. Rio de Janeiro: Editora Nacional/FINEP.

Chacel, J., Pamela S. Falk, and David V. Fleischer. 1988. *Brazil's Economic and Political Future*. Boulder, CO: Westview Press.

Comissão das Sociedades Científicas. 1985. *Ciência e Tecnologia na Nova República: Relatório Apresentado ao Ministério de Ciência e Tecnologia pelas Sociedades Científicas*. Mimeo.

Dahlman, Carl J., and Fonseca, Fernando Valadares. 1987. "From Technological Dependency to Technological Development: The Case of Usiminas Steelplant in Brazil." See Katz 1987.

Dean, Warren. 1989. "The Green Wave of Coffee: Beginnings of Tropical Agricultural Research in Brazil (1885–1900)." *Hispanic American Historical Review* 69 (1): 91–116.

Fernandes, Ana Maria. 1987. *The Role of the Brazilian Society for the Advancement of Science, 1948–1980*. Ph.D. diss., St. Antony's College, Oxford University.

Frischtak, C. 1986. "Brazil." See Rushing and Brown 1986, 31–70.

Katz, Jorge, ed. 1987. *Technology Generation in Latin America Manufacturing Industries*. New York: St. Martin's Press.

Limogi, Fernando. 1989. "Mentores e Clientelas da Universidade de São Paulo." See Miceli, 1989, 111–87.

Miceli, S., ed. 1989. *História das Ciências Sociais no Brasil*. Vol. 1. São Paulo: Edições Vértice.

Nau, Henry. 1986. "National Policies for High Technology Development and Trade: An International Comparative Assessment," see Rushing and Brown 1986, chap. 2.

Paulinyi, E. et al. 1986. *Indicadores básicos de ciência e tecnologia*. Brasilia, Conselho Nacional de Desenvolvimento Científico e Tecnológico. Mimeo.

Romani, J. Pitangui. 1982. "O Conselho Nacional de Pesquisas e a institucionalização da pesquisa científica no Brasil." See Schwartzman 1982, 137–68.

Rushing, Francis W., and Carole G. Brown. *National Policiers for High Technology Industries: International Comparisons*. Boulder, CO: Westview Special Studies in Science, Technology, and Public Policy.

Schwartzman, S. 1978. "Struggling to be Born: The Scientific Community in Brazil." *Minerva* (London), 16 (4): 545–80.

Schwartzman, S., ed. 1982. *Universidades e Instituições Científicas no Rio de Janeiro*. Brasília: Conselho Nacional de Desenvolviento Científico e Tecnológico.

Schwartzman, S. 1986. "Coming Full-Circle: For a Reappraisal of University Research in Latin America." *Minerva* (London), 34, no. 4 (Winter): 456–76.

Schwartzman, S. 1988a. *Bases do Autoritarismo Brasileiro*. 3d ed. Rio de Janeiro: Editor Campus.

Schwartzman, S. 1988b. "The Power of Technology." *Latin American Research Review* 24:1.

Schwartzman, S. 1988c. "High Technology vs. Self-Reliance: Brazil Enters the Computer Age." See Chacel, Falk, and Fleischer 1988, 67–82.

Schwartzman, S. 1991. *A Space for Science: The Development of the Scientific Community in Brazil.* University Park: Pennsylvania State University Press.

Schwartzman, S., and Maria Helena M. Castro. 1984. "Nacionalismo, Iniciativa Privada e o Papel da Pesquisa Tecnológica no Desenvolvimento Industrial: Os Primórdios de um Debate." *Dados: Revista de Ciencias Sociais* (Rio de Janeiro, IUPERJ), 27 (1):89–111.

Schwartzman, S., and Fábio W. Reis. 1989. "Ciência e Tecnologia na Nova República." (Document prepared for the Commission of Scientific Societies.) *Ciência Hoje* 9, no. 50 (Feb.), 62–69.

Germany: Three Models of Interaction—Weimar, Nazi, Federal Republic

Frank R. Pfetsch

The study of science policy in modern Germany is necessarily a comparative one, as a result of the various regimes that have governed Germany in this century. This fact is of particular advantage in examining the relationship between scientists and the state. Each of the three regimes that have existed had their own history of continuity and change. The most discussed issues in this context are:

- How radically did the Weimar Republic break with the regime of the former German Empire?
- Was the Nazi movement a revolutionary one and its takeover of power a revolution?
- Did the Federal Republic of Germany start from "day zero" with new institutions or was it a restoration or renovation of older patterns?

These issues lead to the bases of continuity and change in German history and the role that scientists have played during the various regimes. How did the change of regimes affect the scientific community, and how did scientists react to these changes? Before considering such questions, I propose a framework for analysis.

Models of Interaction

The spectrum of possible types of relationships between scientists and politicians can be seen on a continuum extending from the complete dominance of politics, on the one hand, to the dominance of scientific expertise, on the other. In a setting characterized by the supremacy of politics, decision making ignores scientific advice (*decisionistic model*); hence science has to serve the legitimation of a political elite and the consolidation of its regime. In Germany, the supremacy of the politics model is associated with the tradition of

Carl Schmitt. For the supremacy of science exist two models, one based on basic science or philosophy in the tradition of Plato (*Platonist model*) and the other on applied science in the tradition of Gehlen or Schelsky or in the perspective of postindustrial theories (*technocratic model*); in the latter, experts or scientists become the actual decision makers.

Between these extremes are various possible patterns of interacting and cooperative relationships. These include

- a critical relationship between both sides in which science keeps a distance to politics (*Kantian model*)
- a strict division between both spheres with no connection between them (*Weberian model*)
- a cooperative and communicative relationship between political authorities and scientists in which the two sides interact productively (*pragmatistic model* in Habermas's terms).

Free exchange of communication is hindered by commercial secrecy and the competition between private enterprises, military secrecy in the political competition between nation states, and increasing bureaucratic impermeability as a result of the nature of large-scale research institutes and the ever increasing flood of information.

The mode of interaction depends particularly on the character of the political regime, as determined by the power structures, and on the social organization of the scientific community. Fundamental to my analysis is that the role that scientists play is determined by the structural and ideological orientation of the political regime and the scientific community.

Subsequently I will identify more concretely the kind of relationship that has existed in the very different German regimes of the Weimar Republic (WR; 1919–32), the Nazi Reich (1933–45), and the Federal Republic of Germany (FRG; since 1949). (This analysis excludes a consideration of the former German Democratic Republic.) I will firstly characterize the German scientific community, the physicists in particular, and then main features and orientations of the three regimes with the respective scientific communities and their links to politics.

The German Scientific Community

Before examining the particular role of scientists in the three regimes it is necessary to make some general observations about the German scientific community. At least four characteristics distinguish it from the communities in other countries. One characteristic of German academia is certainly the fact that science in modern Germany, as far as the institutions of higher learning and other research organizations are concerned, has been a public not a private enterprise; thus the organizational schemes of these institutions show the same pattern as that of other areas of public administration. For the most part the people working there were civil servants.

A second characteristic of the German academic community is the idealistic tradition, incorporated in the Humboldtian concept of the university in which the philosophical faculty plays the leading role. With further internal advances (new theories, new inventions) and external advances (new inventions, innovations) in the natural and technical sciences during the nineteenth century, the Germany scientific community split into two different sociocultural worlds, "two cultures" that developed side by side. Institutionally, the two cultures developed within the universities and polytechnics; scientifically, the distinction between *Naturwissenschaften* and *Geisteswissenschaften* marked the two cultures.[1] The attitude of the "German mandarins" (Ringer 1969), their ideology and value systems, stemmed from academic humanists of the nineteenth century and is based on the principles of "pure" freedom of learning and teaching, personal cultivation, and idealistic philosophy; status results from educational qualifications rather than from hereditary privilege or wealth. The self image was that of a "culture-bearing" aristocratic elite (*Kulturträger*). It is important to note in this context that most of the interpretive concepts for the behavior of scientists (such as that mentioned previously) refer to the humanities and not to the natural and technical sciences. The quite distinctive behavior of those in the "hard sciences" reflects their much closer contact with industry and governmental agencies. This must be taken into consideration in science policies in various political regimes.

Third, it should be mentioned that each new beginning in German history had an important impact on scientific development. The loss of the wars in 1808–9, 1919, and 1945, as well as the period after German unification in 1871, each meant a new start in scientific endeavors. Against military defeat and political failure, science was considered a substitute area for German achievements. Science should accomplish what politicians and the military had failed to accomplish for the glory of Germany. Scientific preeminence would thus compensate for national prestige lost in other fields.

A fourth distinctive characteristic of the scientific community as a whole has been the preservation of professional autonomy against external demands or threats. The extent to which it has succeeded in the various political systems is of primary importance to our analysis.

Distinctive Features of the Physicists' Community

What role have physicists played as representatives of their discipline with regard to political issues? How and to what extent has the behavior of the

1. See Brookman 1979, 275–83.

physicists' community reflected or deviated from the general pattern described above?

As other disciplines in the natural sciences such as chemistry, the discipline of physics differs from other fields of research by its importance in production and military purposes. The production character, in particular, provides physics and chemistry a preeminent position within the scientific community (Schwandt 1987). With the development of nuclear power, both for energy production and for military purposes, the previously theory-oriented discipline of physics came to be seen as determining the face of the century (*Jahrhundertwissenschaft*). On the whole, the relatively small share of the physicists' community—with about 5 percent of the university manpower and about 10 percent of the expenditures by the German Research Association (DFG) (see tables 1 and 2)—was by far compensated for by its important role in science policy. As one would expect, physics (as well as chemistry) has played a key role in industry and politics. I will demonstrate subsequently this special relationship of physicists to industry and politics by sketching their role in science policy in general, and, more particularly, with regard to a number of specific projects such as the development of the nuclear bomb, optical instruments, and nuclear energy. As a consequence of the important economic and military roles that both physics and chemistry have played, those in these disciplines have become prominent figures in German science policy in all three regimes.

The Three Regimes: Weimar, Berlin, and Bonn

The three regimes that have existed during the period since World War I are described here in terms of their political, economical, military, and cultural features. It is virtually impossible to characterize the political systems in static

TABLE 1. Personnel in Physics in German Universities

Year	Total Personnel	Number of Physicists	% of Total	Number of Natural Scientists	% of Total
1976	88,673			19,823	22.36
1978	93,249			21,307	22.85
1979	94,403			21,504	22.78
1980	85,234	4,007	4.70	17,474	20.50
1982	89,603	4,371	4.88	19,516	21.78
1984	91,373	4,561	5.99	20,392	22.32
1985	92,916	4,474	5.82	20,500	22.06
1986	96,054	4,698	5.89	21,749	22.64
1987	98,800	4,857	4.92	22,765	23.04

Sources: Statistische Jahrbücher der Bundesrepublik Deutschland 1976–87.

terms because each of them was founded on special historical conditions, and each developed over time. Generalizations are therefore not always justifiable. Nevertheless, one can identify certain specific regime characteristics as well as various phases in the development of each.

The Weimar Republic

Initially the leading politicians of the WR installed a center-left regime with a strong social-democratic orientation. The social structure, however, continued to resemble that of the earlier era, with a number of features that resisted integration into the new constitution. The political system that evolved was pluralistic, if not fragmented, with ever stronger antagonisms among the various political parties and pressure groups. The constitution lacked the support of important strata in the society, and the legitimacy of the regime was questioned by antidemocratic groups and parties. In the end, the *Reichspräsident* was forced to rule with emergency laws. Powerful right-wing groups waged a campaign of violence against the "system," and the Nazis finally fought their way to power.

From its beginning after the defeat in World War I the WR was subject to

TABLE 2. Expenditures for Physics by the German Research Association

Year	Absolute (in mill. RM/DM)	% of Total
1928–33	2.9	11.0
1937	0.6	8.3
1938	0.5	7.0
1939	0.5	6.6
1940	0.3	5.1
1974	55.6	9.0
1975	48.2	9.0
1976	47.4	8.7
1977	59.9	8.0
1978	55.5	8.5
1979	53.4	7.1
1980	65.0	8.0
1982	59.1	7.3
1983	73.6	9.1
1984	78.1	8.7
1985	73.5	7.5
1986	73.2	7.4

Sources: Zierold 1968, 186, 223; Bundesbericht Forschung 1988: Faktenbericht zum Bundesbericht Forschung 1981.

the political influence of other nations, particularly France, and cannot be seen as a fully autonomous state. In time, the politicians tried to play a role between East (Soviet Union: Rapallo) and West (France: Locarno) in order to increase its independence from Western financial and territorial demands. As a result of the defeat and the restrictions imposed upon the new state by the victorious countries, the WR was not involved in any military conflict outside its territory. However, internal crises were widespread; i.e. revolutionary uprising at its beginning in November 1919, passive resistance in the occupied Rhineland, and demonstrations, strikes, and intermittent violence by extremists. The economic history of the WR began and ended in crises due to inflation, reparations, unemployment, and social upheavals. The nation regained the prewar standard of living only in 1927. Due to the unstable economic conditions, the level of exports never reached the prewar level, with exports accounting for 17.5 percent of GNP; in the period 1925–29 it reached its peak of only 14.9 percent.

However, an important fact for the scientific community was that the magnates of the coal and steel industries had profitted immensely from the war and were able to contribute substantial donations to the research institutes serving their interests. The economic situation in the aftermath of war and revolution had a great impact on the personal welfare of German scientists. In the year following the armistice in 1919, the value of the currency declined to less than 10 percent of its prewar level and remained at this level until summer 1921. Then began the period of hyperinflation; this affected mainly the middle class, to which most scientists belonged. The purchasing power of professional income remained far below its prewar levels, and the budgets of research institutes fell even lower. Despite this economic plight, new initiatives were taken by scientists and politicians to institutionalize new research-promoting bodies; this included the Emergency Society for German Science and Scholarship (*Notgemeinschaft der Deutschen Wissenschaft,* NGW); the Emergency Federation for German Science (*Stifterverband der Notgemeinschaft der Deutschen Wissenschaft*), which was a fund-raising foundation from German industry; and others, such as the Justus von Liebig Foundation, Emil Fischer Foundation, Adolf von Baeyer Foundation, and, in particular, the Helmholtz Foundation. In addition, new research institutions were founded by the Kaiser Wilhelm Society. Besides these institutional innovations, other factors must be mentioned in order to explain that during the WR Germany was the home of the "new physics" and attracted many foreign students. Already, during the Empire, the Prussian Ministry with Karl Althoff together with Felix Klein planned for Göttingen to become a center of modern science; public as well as private funding (rather the exception in the German context) supported the university. Together with a skillful appointment system, this policy of concentration (*Schwerpunktbildung*) bore fruit in the years of the WR.

Compared with the FRG, the WR at its beginning was in a somewhat less advantageous situation. The threat of isolation from the international flow of knowledge came from two sources: the lack of foreign currency that affected the supply of literature, instruments, etc., and the boycott declared by leading scientists of former enemy countries against German and Austrian scholars and scientists. They were excluded, for example, from international organizations such as the International Research Council and the International Academic Unions (Schroeder-Gudehus 1972, 551). Since national rivalries played an important role before and during World War I, the postwar period was still marked by nationalist revanchism, which impeded Germany's reentry in the international community.

As to the effect of the regime upon the scientific community, the pluralistic society of the WR allowed a rather free development of scientific endeavors; this was restricted, however, by financial difficulties in the early years of its existence (inflation, reparations), and at its end (the Great Depression). As a result, emergency programs such as those of the Notgemeinschaft der Deutschen Wissenschaft, and the Stifterverand der Notgemeinschaft der Deutschen Wissenschaft, as institutional innovations for raising of funds sought to compensate for these financial problems. An innovation was the allocation of research funds to individual investigators for specific projects on the basis of merit through a review by peers rather than, as formerly, to the directors of institutes. Table 3 shows that over 80 percent of the funds consistently came from the central government. On the whole, the tradition of autonomy for scientific institutions was observed.

Most members of the scientific community opposed the new regime or

TABLE 3. The Financing of the German Emergency Association (in 1,000 RM)

	Government Sources		Private Sources		Total
Years	absolute	in %	absolute	in %	absolute
1920–21	40,000	83.5	7,900	16.5	47,900
1924	3,000	86.2	480	13.8	3,480
1927	8,000	98.0	160	2.0	8,160
1928	8,000	97.4	212	2.6	8,212
1929	7,000	96.3	272	3.7	7,272
1930	7,000	94.3	420	5.7	7,420
1931	5,100	89.4	602	10.6	5,702
1932	4,400	88.2	590	11.8	4,990

Source: Zierold 1968, 38.

were indifferent toward the Weimar "system." Only a minority of prorepublican scholars (*republikfreundliche Hochschullehrer*) supported the center-left regime. Scientists, like those in the military, tried to maintain their independence from government interference and preserve their traditional autonomy.

This behavior corresponds to the Weberian model of noninterference or strict separation between science and politics. The system of self-governing or autonomous institutions therefore also had a political meaning: opposition to the political regime went hand in hand with the principles of autonomy and self-government.

In contrast to the period of Nazi rule, the periods of the WR and the FRG show a relatively open system for the flow of material and nonmaterial goods and ideas. The traditions of internationalism and liberal cultural policy have prevailed in both republican regimes. In the WR, Göttingen, Berlin, and Munich became world centers for physics, but, because of the theoretical orientation of physics, their importance was restricted to academic circles. As a result of this and also of the prevailing apolitical tradition of German academia, the most important self-governing organization for the promotion of science, the *Notgemeinschaft,* showed in the composition of its personnel the classical pattern of German science policy: its board of directors was composed of a representative of the Prussian state (Friedrich Schmidt-Ott), two members of the natural science community (Walter von Dyck, Fritz Haber), and one representative of the humanities (Adolf von Harnack). For the promotion of physical-technical research, the Helmholtz Foundation was more important. Two-thirds of funds collected from industry were to be distributed to the Helmholtz Foundation and one third to the Notgemeinschaft; the formula was later changed to 50 percent for each. This explains why the Notgemeinschaft received so little from private sources. The distribution of grants made by the Notgemeinschaft from 1928 to 1933 to the major disciplines shows almost equal shares to the natural sciences (physics) and the humanities.

The Nazi Regime

The Nazi system also went through changes, and it was never the monolithic totality, that is assumed by some theories of totalitarianism. Instead, the political structure was polyarchic and was based on the four pillars of party, army, industry, and the *Schutzstaffel* (SS); these were bound together by the *führer* and the ideology of racial community (*Volksgemeinschaft*). The instruments used were those of populist charismatic appeal, demagoguery, and physical threat against opposing groups and individuals. From the beginning the system was centralized, with the formerly self-governing states (*Länder*) losing their traditional independence. A "dual state" (*Fraenkel*) emerged, with the centralized administrative system, on the one hand, and the ideologically oriented party, on the other. This pattern of multiple-power blocks allowed rivalry between the blocks and the enforcement of arbitrary authority by the *führer* and those close to him.

The rather reactive and integrative nature of WR foreign policy shifted to the other extreme with the Nazis' aggressive policy toward neighboring countries. Withdrawal from the League of Nations, revindications as to the ethnic German populations in Eastern European countries, and the coerced Munich Agreement show the beginnings of Hitler's expansionist racial policy and his attempt to dominate especially Eastern European countries first by threat and then, from 1939, by war.

From the beginning of Nazi rule, government programs for military and infrastructural purposes brought considerable economic growth and employment. By 1936 full employment was achieved. The political philosophy of self-sufficiency, or even autarky, together with militarization led to increasing centralization of the economy and brought the export/GNP level down to 6.0 percent in 1935–38.

Seen against this background, existing institutions of science policy changed, and new institutions were founded. For the first time in German history, with the establishment of the Ministry for Education, Science, and Cultural Affairs in 1933, a central body for policy-making was created, and in 1937, with the German Research Council, a central tool for research policy.

The pattern of liberal science policy changed considerably during the Nazi regime. Centralization of state policy replaced federalism, and central government institutions were created at the expense of the states. The Ministry for Education, Science, and Cultural Affairs was established, which later founded the Research Council. The main science-promoting institutions, the Notgemeinschaft and, at a later stage, the Research Council were restructured or newly instituted.

The racial and totalitarian policy of Nazi rule opposed both Western (liberal democratic) and Eastern (bolshevist) ideology. As a consequence, the twelve years from 1933 to 1945 were dominated by a revisionistic national policy and the struggle for external domination and internal repression. The partial imposition of a "German science" ideology disrupted international networks and the free flow of ideas.

From 1933 on, the regime's racial policy was applied to the civil service, hence also to university staff. As a consequence, many Jewish scientists were forced to leave Germany. Although accurate figures are not yet available, estimates show that over 1,000 university teachers from all fields had been driven from their posts in 1932–33, or about 14 percent of the staff of German institutions of higher learning. According to Edward Y. Hartshorn (1937) the total was 1,145 university teachers; a later analysis estimated the number to be

1,268 for the twelve-year period of Nazi rule. A breakdown according to scientific fields shows 522 teachers in the Geisteswissenschaften, 374 in medicine, 253 in natural sciences, 75 in technical fields, and 44 in theology.[2] Among those who departed from posts in Germany between 1933 and 1945 were 11 Nobel laureates in physics, 4 in chemistry, and 5 in medicine. The figures differ from university to university. Those suffering the greatest losses were the liberal universities of Göttingen, Berlin, and Heidelberg, losing, respectively, 53, 68(?), and 62 appointed teachers (full professors, associate professors, and instructors; i.e., persons with *Habilitation*), or about 30 percent of the teaching staff.

The image of persecuted scientists has, however, many facets. The majority of university teachers did not oppose the new regime and tried to preserve their traditional autonomy or at least independence; some sympathized with the Nazi "revolution." Severely hurt by Nazi rule were certainly those who opposed the regime and those of Jewish decent or married to Jews. At the end of the Nazi rule, there was practically no teacher of Jewish descent left in German universities.

This picture, however, did not remain the same for the whole period of Nazi rule. More particularly, the situation of the physicists' community changed drastically during the period of Nazi rule. In 1934, the Nazi physicist Johannes Stark, Nobel laureate and president of the Imperial Institute of Physics and Technology, became president of the Notgemeinschaft. Other physicists—some of them also Nobel laureates—who were of Jewish decent, such as Einstein, Born, Franck, and Teller had to leave Germany.

With the first Four Year Plan from 1936 on, the hostility toward science and the dominance of ideology over scientific standards and achievements abated. With the policy of preparation for war, science and technology again assumed an important role. Links between the natural sciences, engineering and industry became much closer and because of its pragmatic value a set of scientific standards increasingly prevailed. Hence, scientists became much more respected agents in the process of science policy decision making; this is demonstrated by an analysis of appointments in the various institutions. For instance, the Nazi physicist Stark was removed from the Imperial Institute for Physics and Technology and the Notgemeinschaft, and not appointed for the "Uranium Project" during the war (see subsequently).

Relations between science and politics in this period must be regarded from quite different angles. Three questions may be asked. What influence did the totalitarian regime exercise upon the scientific community? How did Nazi ideology affect scientific orientations? How did science help the regime?

Regarding the first of these questions the leading Nazi politicians did not

2. See Stifterverband für die Deutsche Wissenschaft 1950, 18; and Hartshorne 1937, 93.

consider the discipline of physics to be worthy of special treatment. Although Stark was aware of its special importance for German economic independence in the search for substitutes to imported raw materials, Hitler and the other Nazi leaders did not realize the importance of science for production. Only during the war did its military importance become evident, and natural scientists then enjoyed a somewhat better life and were no longer subjected to discrimination. In his analysis of the scientific community during the Hitler regime Alan D. Beyerchen (1977) takes Göttingen as an example to demonstrate that only one-third of physicists and mathematicians remained at their jobs. With the "Law for the Restoration of the Civil Service" in 1933, non-Aryan public employees as well as those who did not support the new regime could be dismissed unless they served in the armed forces, "fighters on the front," during World War I. James Franck, an exsoldier himself, was one of the few who protested publicly. His colleagues, Max Born and Richard Courant, left Germany. In 1935, almost one-fifth of all natural scientists had left their positions. Albert Einstein and Fritz Haber were among the most prominent of these. Only a few natural scientists actually sympathized with the new regime, but the majority hoped that Hitler would be able to solve the economic crisis and that the "brown tempest" would then disappear. Among the community of the natural scientists only Robert Havemann is known to have actively opposed the regime.

In some cases, though, threatened natural scientists were able to continue working in Germany until the end of the war. The case of the Zeiss glass and optical industry in Jena is illustrative. Some researchers were employed there until the end of the war even though they were of Jewish decent (e.g., Otto Eppenstein), married to Jews (e.g., Hugo Schrade, Rudolf Straubel), or active members of previous opposition parties (e.g., Friedrich Schomerus). The reason for this lay in their prominent role in a militarily important industry.

Karl-Heinz Ludwig (1974) has shown that due to their incompetence, rivalry, and intrigues, politicians and politically minded scientists were not able to impose Nazi ideology upon the scientific community. Also, the most important instrument of research policy by the Ministry for Education, Science, and Cultural Affairs and the Research Council that was founded in 1937 failed in controlling science and technology. The jurisdiction of these bodies did not extend to industrial research, and the chemical industry was incorporated in the Four Year Plan (1936–40); military research was conducted by the three military branches. It was the Armament Agency (*Reichswaffenamt*) that was responsible for the development of nuclear fission (see subsequently).

The indoctrination or penetration of the scientific community with Nazi ideology—the second question—succeeded in the case of only a few natural scientists. At the beginning of World War II, for example, only six out of eighty-one chairs in physics were held by "German physicists." The most prominent among them were the Nobel laureates in physics Philipp Lenard and Johannes Stark. Their "German physics" was attacked by others in the discipline, and in 1940 the dispute—which also was an internal dispute between two schools of thinking—was decided by the Nazi party in favor of modern theoretical physics. Professional autonomy prevailed over ideology.

The third question, that of support for the regime among scientists, must be seen in the proper time frame. In the beginning, ideology prevailed over standards of the discipline. The "revolution" was the vehicle for transporting Nazis and their supporters into official positions. Nazi scientists became heads of important organizations. Other bodies in charge of science policy, such as the Federation of Foundations and the Kaiser Wilhelm Society, remained in the same hands as before. The preparation for war began in 1937, and, for this reason, scientists and engineers rose in importance regardless of their political or ideological inclinations. In this period a closer alliance among natural scientists, engineers, and industry can be observed, thus establishing a technocratic approach in the making of science policy.

Construction of the Nuclear Bomb

An important project during the war was that of developing nuclear fission. In 1939, the German army initiated a program for the utilization of nuclear fission. For this purpose the Kaiser Wilhelm Institute for Physics in Berlin was confiscated. The resulting German "uranium project," headed by Werner Heisenberg, was conducted at about the same time as and—in its later stage—in competition with the Manhattan project in the United States. The team of scientists, including Heisenberg, von Weizsäcker, Bopp, Harteck, Bothe, and others, who worked on the "uranium project" knew from 1942 on (and some even earlier) that the program could lead to the construction of a bomb. The motivation for their work lay principally in scientific curiosity and the desire to control new scientific developments. Their justifications were that the nuclear bomb could not be built before the end of the war, and that the acquired knowledge could be used after the war for peaceful purposes, namely to provide energy. A third argument reported by von Weizsäcker was a moral and political one: the existence of the atomic bomb would, because of its unimaginable destructive capabilities, prevent its employment and end the six thousand-year-old history of war. The development of the bomb would require immense resources and a long period of time.[3] Because of this Albert Speer, the minister responsible for armament matters and the implementation of the second Four Year Plan, decided to reduce the project to a minimum

3. These arguments are reported in a recent interview with Carl Friedrich von Weizsäcker in *Der Spiegel,* no. 17, 1991.

level. The assessment made by Heisenberg and Speer that the investment necessary to realize a working bomb before the war's end was beyond the capacity of the Nazi regime later proved realistic in the light of the much bigger resources available for the Manhattan project. This indicates that there was no genuine race between the German and the U.S. projects.

The Federal Republic of Germany

The constitutional framework of the FRG must be seen in the Weimar tradition, with improved foundations, and filtered through the experience of Nazi injustices. The new system was referred to as "Chancellor democracy" because of its stronger emphasis on coherent executive power based on a party majority in parliament. The interest groups that evolved became an important factor, leading some scholars to speak of a corporate system.

From its founding in 1949 the FRG was integrated into a liberal world economy. With its domestic social market economy of nonplanning and its externally free trade policy, the FRG became part of the Western industrialized system, with its concentration in Western Europe (EEC, EC) and the Atlantic Alliance (NATO). The foreign policy of the FRG was determined by five problem areas; i.e., by the search for unification, security (mainly the protection of West Berlin against the Soviet threat), economic welfare, the burden of the past, and the antagonistic international environment of the Cold War. In close cooperation with France, the first Chancellor, Konrad Adenauer, gave priority to Western European integration over national unification. The European policy was a constitutive part of at least three of the five most important elements of national interest; i.e., the unification problem, with its external instrument of the so-called Hallstein Doctrine, and the economic necessities of export substituted growth. As a third element, the lessons of the past made nationalism obsolete and Europe became its substitute. As a result, Germany as a federal system itself was, and is, more open to European cooperation and favors a federative type of integration. As such, the FRG was so far not involved in armed conflicts. The unification on 3 October 1990 concentrated much of its resources to the rebuilding of the five new Länder of the former Communist eastern state. Scientific institutions and science policy organisms became reorganized according to the "Western model".

Economically the recovery of the FRG in the 1950s led to an increasing integration in the world market. This is demonstrated by the increase in the share of exports in the GNP valued at market prices: 9.3 percent in 1950; 17.2 percent in 1960; 23.8 percent in 1970; and 26.7 percent in 1980. Its export-supported economic model geared the country toward integration into the world market to a degree that it had never known before.

The period after World War II was, due to the U.S. "open door policy," much more receptive to the German scientific community. New science policy-making and research institutions were founded. In 1956 the Ministry of Nuclear Energy was created for the purpose of developing nuclear energy and administering nuclear power stations. Later the role of this ministry was expanded, and it became a body of the central government for the promotion of various R&D programs.

The FRG started with a more decentralized pattern of federalism, which in time evolved toward increasing responsibility on the part of the federal government. Since the mid-1950s, the federal government assumed the responsibility for nuclear energy, and, in the following years, for other research fields of national interest such as information science, space, and environment, as well as military R&D. Since 1969, federal ministries have expanded their role within the institutes of higher education by participating (together with the Länder) in the financing of university infrastructure and by formulating legal responsibilities. The main sources of funding R&D, however, have come from industry, and this has consequently been the chief beneficiary (see fig. 1 and table 4). The attitude of the scientific community has been basically positive toward the FRG. Especially natural scientists have been able to influence governmental science policy.

Although some issues such as remilitarization, deployment of nuclear weapons, and, more recently, increased promotion of nuclear energy have been controversial and have found scientific spokesmen in public debates, the intellectual climate on the whole has been favorable toward science and has made scientific endeavor a worthwhile occupation. The early history of the FRG, up to the 1960s, saw material and intellectual shortages and despair. In 1945–65 about 5,000 technicians and scientists left Germany for better working conditions abroad, primarily in the United States. At the end of the 1960s, a program of recuperation tried to win them back.

Politically, the scientific community was represented by a cartel of three institutions, the German Research Association (DFG), the Max Planck Society (MPG), and the West German Conference of University Rectors (WRK).

Due to the restrictions imposed by the victorious countries after World War I and II, Germany showed a relatively low level of military spending, and, as a consequence, low figures for military-science expenditures compared with other European countries (see table 5). During the era of the WR military-science expenditures were almost nil, and, for the Nazi period, I have calculated that this portion of the budget did not exceed 3 percent. Compared with other western countries, such as the United States, Great Britain, or France, the FRG had very low science expenditures in the military field. In 1977, for example, they amounted to 12.7 percent of the federal government's expenditures, as compared to 6.2 percent for economic purposes and 47.5 percent for the advancement of science in general (see table 6). Taking into

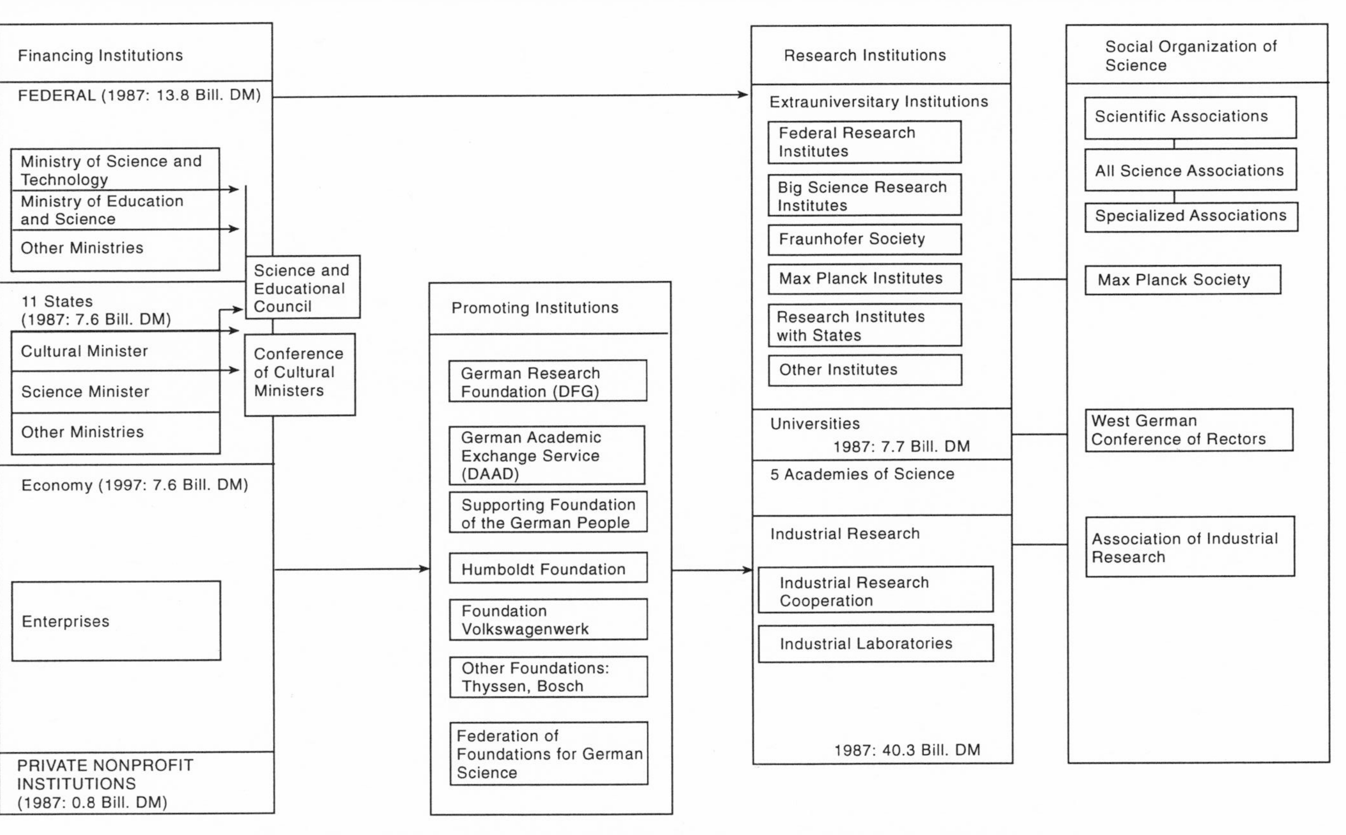

Fig. 1. Organization of the science system of the Federal Republic of Germany

TABLE 4. R&D Funding in Germany 1925–87 (in Mil. RM/ DM)

			Governmental				
	Private Industry		Academic		Nonacademic		
Year	absolute	in %	absolute	in %	absolute	in %	Total
1925			16		94		
1930			10		95		
1935			19		100		
1940			51		179		
1945			—		—		
1950			0.1		124		
1955			0.1		184		
1960			0.2		627		
1965	4,060	51.33	290	3.67	1,790	22.63	7,910
1970	7,610	51.42	1,152	7.78	4,010	27.10	14,800
1975	11,792	47.85	1,382	5.61	7,410	30.07	24,645
1976	12,600	48.95	1,333	5.18	7,408	28.78	25,740
1977	14,109	50.87	1,162	4.19	7,642	27.55	27,735
1978	16,870	53.35	1,202	3.80	8,473	26.80	31,620
1979	18,663	54.13	1,063	3.08	10,033	29.10	34,477
1980	19,895	54.30	1,081	2.95	10,468	28.57	36,641
1981	22,082	56.12	1,037	2.64	10,784	27.41	39,345
1982	23,560	55.92	1,122	2.66	11,978	28.43	43,942
1983	25,459	57.94	1,444	3.29	11,751	26.74	43,942
1984	26,990	58.62	1,392	3.02	12,104	26.29	46,040
1985	31,093	60.26	1,240	2.40	13,253	25.69	51,598
1986	32,700	61.10	1,387	2.59	13,358	24.96	53,516
1987	34,700	61.03	1,280	2.25	14,412	25.35	56,860

Sources: Datenhandbuch (1985), 118ff.; Bundesbericht Forschung (1988), 351, 354; Faktenbericht zum Bundesbericht Forschung (1981), S. 366.

account total governmental budgets (federal as well as Länder), the share of the science expenditures for military purposes falls to only 5.1 percent as compared with 64.5 percent for the advancement of science.

Compared to other Western countries, Germany today shows a civilian orientation to R&D; the United States, United Kingdom, and to a lesser extent, France show a military orientation; Sweden and Japan strongly promote industrial development.

In 1957–58 the deployment of U.S. missiles on German soil became a public issue. Many scientists then became politicized, opposing the official policy. An alignment of various groups and organizations opposed the stationing of atomic weapons in the FRG, and these issued the "Statement of the Göttingen Eighteen," with prominent physicists among them.

The knowledge acquired since the 1920s and especially during the war in the "uranium project" was able to be used for nuclear energy production after

TABLE 5. Science Expenditures for Military Purposes as a Percentage of Total Science Expenditures

Year	Military Science as % of Total Science Expenditures
1925	0.1
1930	0.4
1935	2.9
1955	0.1
1960	6.7
1965	9.0
1970	8.4
1975	5.1
1980	—
1985	12.5

Sources: Pfetsch 1985; for 1985 R&D expenditures, not science expenditures, according to Bundesbericht Forschung 1984.

the war. The supply of energy was seen as a crucial factor in postwar industrial development in the FRG. It was thought that the coal reserves in the Ruhr and Saar regions would soon be exhausted, and that a substitute would be needed. Nuclear power was seen as an ideal alternative, and physicists began to be interested in the field. From the mid-1950s, the FRG established an administrative framework with a federal ministry and links to industry and Länder governments, leading to the foundation of research centers in Karlsruhe and Jülich. The federal government thereby assumed responsibility for a field that was previously in the juridiction of the states. Later, in the 1970s, with the public-action interest groups and the peace movement, forerunners of the Green Party, physicists such as Traube played a public role in opposing the future construction and use of nuclear reactors.

TABLE 6. R&D Expenditures of Various Countries by Major Aims (central governments)

Country	Defense	Energy	Space	Industrial Development	Advancement of Science
U.S. (1976)	49.6	9.6	14.8	1.5	3.8
U.K. (1975)	48.4	6.5	0.7	10.3	21.4
France (1977)	30.6	8.4	8.4	11.4	25.2
Germany (1977)	12.7	12.3	6.8	6.2	47.5
Sweden (1975)	11.5	3.8	0.9	53.8	23.4
Japan (1975)	3.7	16.6	1.6	21.5	5.2

Source: Unesco, *Unesco Statistical Yearbook*, 1987.

Another dimension that should be mentioned is the coordination of science policy within the framework of the European Community.

Compared with the period after World War I, the situation for the scientific community after World War II was considerably better. This was due to the openness to the international community, free access to Western information sources, links to émigré scientists (some of whom returned to Germany), together with economic recovery programs that had not existed to this extent after 1919. However, destruction from the war, losses from emigration, political insecurity, etc. made the new beginning difficult. As a result, scientists looked for better working conditions elsewhere, and some found them, mainly in the United States. From 1949 to 1963, the brain drain brought 54,558 technicians and natural scientists to the United States, and the proportion of German scientists and technicians among them was 8 to 10 percent.[4]

Some of the scientists who remained in the FRG obtained key positions. Representatives of the most influential science policy organizations (DFG, MPG, and WRK) came principally from physics; among them was Carl Friedrich von Weizsäcker, a prominent spokesman on political issues such as rearmament, deployment of nuclear missiles, and the utilization of nuclear energy. A few physicists such as Traube became prominent environmental protection campaigners.

Also, the new type of big science institutions, those in Jülich, Karlsruhe, and Hamburg, which were founded in the late 1950s and 1960s had physicists as founders or promoters. The federal government with the establishment of the Ministry for Nuclear Energy, which grew continuously and changed its name several times, assumed responsibility in matters of science and technology. The main project, that of nuclear fission, was in the hands of physicists. The main advisory body, the Nuclear Energy Commission, boasted prominent members from the physics profession.

The unification of East and West Germany was a process by which the new Länder became part of the old ones. This means that all of the important science and science policy institutions extended their frame of action into the East with the effect that universities expanded substantially and the former academies were reduced in size and, after successful evaluation, became part of western organizations.

Conclusions

The history of science policy in Germany shows both long-term traditions and short-term contingencies. Traditions that prevailed throughout the various

4. See: Müller-Daehn 1967, 71. His estimates show a total of 5,604 immigrated German technicians and natural scientists for the period 1945–65, among whom were 3,941 technicians and 1,663 natural scientists (see 73).

periods are those related to certain basic features of the German scientific community. Those include the promotion of state institutions, idealistic approaches to scientific endeavors, and attempts to preserve independence from the state and from industry. Occasional contingencies are those due to changes in the political, economic, and military fields that have affected scientists either positively or negatively.

In portraying the relationship between politics and science in each of the three regimes the following characteristics may be noted. The period of the WR, because of its conservative scientific community that largely opposed the WR, can be characterized by a relationship of opposition, passive indifference, or pragmatic accommodation when its own interests were at stake. The most appropriate model for describing these relations is that of Weberian strict noninterference or separation.

During the Nazi period, in contrast, the relationship on the part of most scientists was one of accommodation. Only a minority of opponents and the racially discriminated had to chose internal or external emigration, hostility, or passive resistance. The most appropriate model for describing this behavior is that of a decisionist stance on the part of politicians in the early years of the regime and that of the technocratic model in the case of physicists for the later years. Also, accommodation and, in a few cases, deadly encounters on the part of certain sectors of the scientific community can be observed.

For the FRG the relationship between science and politics can be seen largely as one of "happy convergence" of interests or indifference in the apolitical tradition of German idealistic philosophy. The basic consensus between the Bonn governments and the scientists has never been questioned, even during periods of partial confrontation. The most appropriate model for describing relations in this period is that of noninterference (autonomy) and pragmatistic cooperation on the part of the scientific community, and that of critical distance in the Kantian tradition for a minority.

SELECTED BIBLIOGRAPHY

Abelshauser, Werner. *Wirtschaftsgeschichte der Bundesrepublik Deutschland 1945–1980*. Frankfurt, 1983.

Becker, H.; Dahms, H.-J.; Wegeler, C.; eds. *Die Universität Göttingen unter dem Nationalsozialismus*. Munich, London, NY, 1987.

Beyerchen, Alan D. *Scientists under Hitler: Politics and the Physics Community in the Third Reich*. New Haven and London, 1977.

Brookman, Frits Henry. *The Making of a Science Policy*. Ph.D. diss., Vrije Universiteit, Amsterdam, 1975.

Buselmeier, K.; Harth, D.; Jansen, Ch.; eds. *Auch eine Geschichte der Universität Heidelberg*. Mannheim, 1985.

Forman, Paul. "Weimar Culture, Causality, and Quantum Theory, 1918–1927." *Historical Studies in the Physical Sciences* 3 (1971): 1–115.

Forman, Paul. "The Financial Support and Political Alignment of Physicists in Weimar Germany." *Minerva* 12 (1974): 39–66.

Hartshorne, Eduard Y. *The German Universities and National Socialism.* London, 1937.

Hermann, Armin. *Die Jahrhundertwissenschaft.* Stuttgart, 1977.

Hermann, Armin. *Nur der Name war geblieben.* Stuttgart, 1989.

Ludwig, K. -H. *Technik und Ingenieure im Dritten Reich.* Düsseldorf, 1974.

Lundgreen, P., (ed.) *Wissenschaft im Dritten Reich.* Frankfurt, 1985.

Mehrtens, H., and Richter, S., eds. *Naturwissenschaft, Technik und NS-Ideologie.* Frankfurt, 1980.

Müller-Daehn, Claus. *Abwanderung deutscher Wissenschaftler.* Göttingen, 1967.

Pfetsch, Frank R. *Datenhandbuch zur Wissenschaftsentwicklung: Die staatliche Finanzierung der Wissenschaft in Deutschland 1850–1975.* Zentrum für historische Sozialforschung. Köln, 1985.

Ringer, Fritz. *The Decline of the German Mandarins: The German Academic Community, 1890–1933.* Cambridge, 1969.

Schroeder-Gudehus, Brigitte. "The Argument for the Self-Government and Public Support of Science in the Weimar Republic." *Minerva,* 10, no. 4 (October 1972): 537–70.

Schwandt, Jurgen. *Toward a Theory of the Modern State.* In J. L. Lzylionicz, ed. *Technology and International Affairs,* 43–97. New York, 1981.

Stamm, Thomas. *Zwischen Staat und Selbstverwaltung: Die deutsche Forschung im Wiederaufbau 1945–1965.* Cologne, 1981.

Stifterband für die Deutsch Wissenschaft, ed. *Forschung heißt Arbeit und Brot.* Essen, 1950.

Unesco. *Unesco Statistical Yearbook.* Paris: Unesco, 1987.

Walker, Mark. *German National Socialism and the Quest for Nuclear Power, 1939–1949.* Cambridge, MA, 1989.

Zierold, Kurt. *Forschungsförderung in drei Epochen.* Wiesbaden, 1968.

India: The Nuclear Scientists and the State, the Nehru and Post-Nehru Years

Ashok Kapur

Although the Indian scientific establishment is large in size it is not truly a scientific community sharing a common set of scientific values, a commitment to scientific criteria, or common interests. Indian scientists lack internal coherence, and most members of the science pool are marginalized, depoliticized and noninteractive.[1] Despite these structural systemic characteristics there is a history of extensive policy driven interaction between scientists and politicians at the top of the Indian policy pyramid. The major forces shaping this interaction were power struggles among key decision makers, bureaucratic politics regarding the goals and strategies of Indian policies, and changes in international relations affecting the definition of Indian-national interests. These external influences often radicalize Indian political processes and, in turn, increase domestic pressures on the decision makers.

India's science policy rests on a number of principles.

1. There must be secrecy in science policy-making, resource allocation, and science administration, but the main aims of science policy must be legitimized by the Indian Parliament.
2. The organization of scientific activities in India must be compartmentalized along functional lines; the habit of working on a need-to-know basis is institutionalized.
3. Dual-use, civil or peaceful-military, scientific and technical activities are preferable given the need to acquire options to enable the Indian government to deal with a variety of internal developmental, diplomatic, and security contingencies.

This chapter draws on research supported by the Social Sciences and Humanities Research Council of Canada.

1. For problems in Indian science planning see V. Shiva and J. Bandyopadhyay, "The Large and Fragile Community of Scientists in India," *Minerva* 18, no. 4 (1980): 575–94.

The key players in the newly independent India were the scientific and political fathers of Indian science before 1947, who emerged as the czars of Indian science policy after 1947. The majority of workers and administrators in the scientific establishments play only a marginal role. This fact justifies the strong emphasis in my study on scientific and political personalities in the Indian case. Politically conscious nationalist scientists in pre-independence India established coalitions with politicians, especially J. L. Nehru, before 1947.[2] After independence (1947), the scientists in the dominant pre-1947 coalition captured high government positions in India. They set the research agenda as well as helped define the legislative and institutional framework of policy. In other words, they captured the state apparatus and created an autonomous scientific empire—a state within the Indian state. J. L. Nehru, Prime Minister of India (1947–64), the minister in charge of atomic energy, and a politically dominant member of the pre-1947 coalition shaped the policy agenda and the legitimacy of scientific activity.[3] The scientists pursued the research agenda without political interference. They received significant and continuous support by the state by way of funds, official sponsorship, and political legitimacy. Scientists inside the policy pyramid intervened and put pressure on the political leadership to change Indian policies. The opportunities to exert such pressures were provided by regional crises, especially the India-China and India-Pakistan wars, and by international developments, especially China's nuclear program and the establishment of the Nonproliferation Treaty (NPT) regime.[4] In general, Indian scientists were able to increase the military content of India's nuclear policy, to enable Indian diplomacy to resist international pressures to denuclearize India, to create the scientific infrastructure to support civilian as well as military applications of nuclear and space science, and generally to shift the orientation of Indian society and the Indian government away from a faith in disarmament and

---

2. R. S. Anderson, *Building Scientific Institutions in India: Saha and Bhabha,* Centre for Developing Area Studies, McGill University, Occasional Papers Series no. 11 (Montreal: McGill University 1975).

3. Nehru was writing about atomic affairs as early as 1933 (see his *Glimpses of World History,* (London, L. Drummond Ltd., 2d ed. 1942, 867–68). Nehru's 1937 address to the Indian Science Congress revealed his emphasis on the link between science (including the atom) and human development. Bhabha lobbied in March 1944 with the Tatas to invest in Indian atomic energy development. Sir J. C. Ghosh, Director of the Indian Institute of Sciences emphasized the applications of atomic energy in India in August 1945. These views emphasized the peaceful uses of atomic energy and science. Nehru's dual-track policy foci—that atomic energy was meant for constructive purposes, but it could be used for defensive purposes if necessary—was outlined on 26 June 1946 (cited in K. K. Pathak, *Nuclear Policy of India,* [New Delhi: Gitanjali Prakashan, 1980], 3, 9, and 10).

4. On the role of external influences on the domestic political economy of science see Etel Solingen (in this volume).

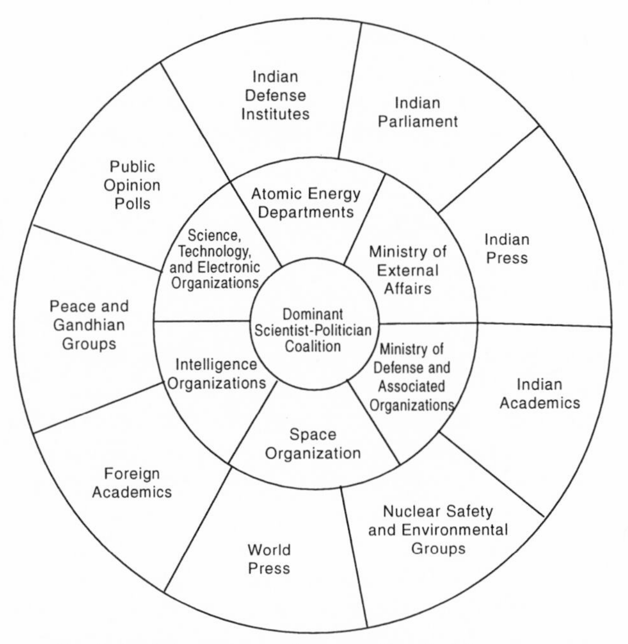

Fig. 1. India's nuclear science policy decision structure

peace, and toward a policy of internal strength and Indian autonomy in international nuclear and scientific affairs.

In my chapter *state* means the dominant coalition, policy network, and bureaucratic politics. Generally, the Indian legislature, academics, the press, and the majority of scientists are not involved in high level intragovernmental controversies. These debates are generally protected from public scrutiny by the stringent Official Secrets Act. However, regional crises, especially the India-China war (1962), China's nuclear test (1964), and the India-Pakistan war (1965), radicalized the Indian political system. These crises led the dominant scientist-politician coalition to resist external pressures and polarized the decision-making process between those who wanted Indian nuclear arms, and those who wanted a denuclearized India.

Indian science policy processes may be visualized in terms of three concentric circles (fig. 1). The inner circle is occupied by the dominant scientist-politician coalition. The second (outward) circle is occupied by the government (political, economic, defence, scientific, and intelligence) bureaucracies at large. It responds to the direction of the inner circle of scientific and political czars, but it also provides inputs into the inner circle especially in crises. The third circle includes the legislature, press, public opinion, and academics and has two main functions: to legitimize public policy by approving actions of the executive branch, and to express public opinion in times of crisis. In other words, the real authority to make policy lies with the first and the second circles. The third circle is generally passive in the decision-making process, but the second and third circles can be significant in crisis situations. Given this structure of policy-making power and science administration, I have chosen to emphasize the role of the scientific and political personalities in the first circle rather than the entire scientific and political apparatus.

Historical Patterns: Happy Convergence and Ritual Confrontations

The first pattern of state-scientists interaction crystallized in the context of a changing state form—i.e., when British colonial rule in India was in the process of being replaced by an independent India. Between the 1930s and the early 1950s a pattern of happy convergence and ritual confrontation evolved around the political aims and the science strategy of the key players—Saha, Bhabha, Bhatnagar, and Mahalabonois (the scientists); and Nehru and Bose (the politicians).[5] Saha, a physicist and a contemporary of Bhabha, disagreed with the latter's approach to India's nuclear science policies and administration. Saha was politically close to Bose, a political leader who sought Indian independence through an armed struggle and who opposed the politicies of Gandhi and Nehru in the Indian independence movement. Bhabha, a close confident of Nehru, became the czar of Indian nuclear science. His approach to develop Indian nuclear science is still the basis of India's policy despite his death in a mysterious aircrash in 1966. Bhatnagar enjoyed the confidence of Bhabha and Nehru and he ran India's scientific establishments during the Nehru years. Mahalabonois, a statistician, organized the Indian Statistical Institute that performed the data analysis for India's five-year plans as well as Indian scientific planning. They all agreed on the need to secure power and self-reliance for India and to use science to solve national development problems. However, their aims and science strategy revealed sharp differences. Saha emphasized the following.

- Complete reorganization of the society and economy of India.
- Rational and national planning with an emphasis on industrialization and the public sector.
- The state must address energy problems including nuclear power and hydroelectric resources.
- The Gandhian approach to economic change (with an emphasis on village industries rather than industrialization) is backward-looking.

5. See Solingen (in this volume) on happy convergence and ritual confrontation.

- Atomic Energy Commission's (AEC's) links with private industry and foreign firms are detrimental.
- Universities should be the foundation of research in India. The Council for Scientific and Industrial Research (CSIR) approach to create a chain of research institutes independent of universities is undesirable.
- Science must have a social purpose.

In contrast, Bhabha's agenda emphasized the following.

- There is a need to build and support fundamental research.
- Fundamental physics are the spearhead of research and practical applications in industry. (The USSR was the model in this regard.)
- Nuclear energy is important for energy production in India.
- It is valuable to create research groups around suitable scientists.
- Government *support* for research without government *control* over scientific research is desirable (i.e., an emphasis on the autonomy of science).
- Need to develop international scientific contacts and exchanges to facilitate Indian scientific activities.

Bhatnagar's agenda was linked to Bhabha's and emphasized the creation of a chain of state-funded research institutes independent of Indian universities. The encounter was between the Saha-Bose approach and that of Bhabha, Bhatnagar, and Nehru. A key scientist and a key politician converged in each coalition. The struggle between the two was settled in 1947 with the victory of Bhabha and his allies. The winning strategy would determine who among the Indian scientists would guide science policy in the newly independent India, what method would help organize India's scientific infrastructure, and what the nature of India's nuclear policy should be. By 1947 the first and the second questions were settled in favor of the Bhabha-Bhatnagar line. The third question was addressed in an ambivalent policy stance that stressed India's commitment to develop the peaceful uses of nuclear energy, and India's faith in security through disarmament. However, Prime Minister Nehru's position also included a reference to "defense use of atomic energy, if India is compelled to do so." The winning coalition emphasized the need to harness science for Indian economic development, and the need to mobilize national resources and national power, but downplayed the military dimension of this link. Out of these twin concerns came the strategy to develop dual-use technologies and to develop international scientific cooperation between India and the scientific leaders in the world. This strategy implied seeking technologies and materials with clear civilian applications as well as military uses or spinoffs. Along with this strategy was a policy and posture of peaceful uses only but with a built-in defense use if necessary.

The Indian case thus reveals two patterns of state-scientists relations. First, by 1947 a pattern of happy convergence emerged in Indian scientific affairs. Following Indian independence and Nehru's rise to national power—he dominated Indian foreign and atomic energy affairs after 1947—the pre-1947 controversy between Saha and Bhabha was settled in favor of Bhabha. Bhabha's political patron, before and after 1947, was Nehru, and the two had a special relationship where Bhabha called Nehru "*bhai*" (brother) in his frequent and private meetings with the prime minister. A critical part of the Saha-Bhabha controversy concerned the role of Indian universities as primary centers of scientific work. Saha favored the universities. Bhabha and Bhatnagar (later head of Indian Council for Scientific Research) favored government funding for highly specialized and functional research institutes that depended on government funding but were nevertheless free to choose their research priorities and strategies without interference by politicians and civil servants. The Nehru government agreed to the latter approach. This created a happy convergence in the form of active state-scientists collaboration through the development of research institutes that were affiliated with the government. This gave the establishment scientists a basis of political power as well as financial autonomy and a freedom to pursue their research agendas within the overall framework that sought to link science to national development.

This happy convergence allowed a leap forward for India in a historical perspective. Ancient India had a tradition of quality scientific work in different branches of science followed by centuries of social, economic, and political backwardness and foreign conquest. In the first half of the twentieth century, "Indian" scientific work was limited to a few universities and involved a few Indian scholars. However, the British India government introduced Western scientific ideas and technology through its use of industrial revolution technology such as the steam engine, geological surveys, and the development of functional (sector-oriented) research institutes (see app. A). After independence, the government of India under Nehru's direction embarked on a plan to develop India's scientific infrastructure on modern lines. This was done by developing a network of state-funded and state-directed semiautonomous institutes in a variety of scientific areas such as atomic energy, space, scientific and industrial research, electronics, energy, environment, ocean development, medical research, agricultural research, and biotechnology. Today there are about two hundred research laboratories in these areas. More recently, the aims of scientific activities were outlined in the science policy resolution adopted by the Indian parliament in March 1958 and the technology policy statement released in January 1983. These stressed the importance of state intervention in securing the benefits of science and technology for the Indian people. They stress the importance of self-reliance and social and economic development as national goals.[6]

This collaboration also created a two-tiered research infrastructure in India. The government-affiliated research institutes were the primary beneficiaries of government funding and they had decision-making autonomy regarding scientific affairs, subject to nominal supervision by the prime minister who also held the atomic and science portfolios. The scientists heading these institutes formed the inner circle of Indian science administration. The second tier consisted of isolated and politically weak universities and research centers that received limited government support but were marginal in the decision process; they were outsiders. There was happy convergence between the state and the inner circle of scientists but growing tension between the inner and the outer circle of scientists. Yet this tension did not lead to ritual confrontation because the inner circle enjoyed political power (and access to the prime minister and the relevant government departments), and they controlled the national scientific establishment.

The second pattern was one of emerging ritual confrontation over nuclear policy matters. These encounters had little impact on the happy convergence between scientists and the state, because scientists were tenured once appointed to government posts and to research institutes. They could not be purged and policy differences inside the Indian government remained secret because of India's stringent Official Secrets Act. The participants in these encounters were all insiders, the relevant scientists, politicians, and civil servants.

The pattern of encounters between Indian scientists and the state's political leadership was the result of frequent interactions between nationalist scientists (who held government positions) on the one hand, and ambivalent political leaders, "foreign linked" sections of India's foreign policy establishment, and the scientific bureaucracy on the other. These encounters dampened the search for autonomy and power development in Indian diplomatic and military affairs during the mid-1960s and 1970s, but they did not eliminate it at the time. The main effect of these encounters was to create ambivalence (oscillation) and internal controversies about the military applications of Indian science. However, by the 1980s the nationalist sector of the Indian scientific community and civil bureaucracy gained the upper hand, and other sectors of the Indian bureaucracy and science administration lost the struggle to denuclearize India. As a result, Indian scientists decisively changed the orientation of Nehru's disarmament diplomacy and his peaceful (nonmilitary)

---

6. For details see *India, 1988–89* (Publications Division, Ministry of Information and Broadcasting, Government of India, 1989).

nuclear policy. By the 1980s Indian nuclear and space activities acquired military content.[7]

These nationalist versus foreign linked internationalist confrontations point to the existence of two competitive subcultures in the upper echelons of Indian decision making since 1947. The first one was nationalistic, and was represented by Dr. Homi J. Bhabha and his allies. Its political drive was shaped by attitudes formed prior to Indian independence in 1947 and its influence after 1947 was evident in India's fears of a superpower-dominated international arrangement in nuclear and disarmament affairs. This approach was expressed in the following positions. (1) November 1948—India supported the Baruch Plan but opposed international ownership of uranium and thorium. India possessed thorium and recognized it as an economic asset. (2) May 1954—in commenting on President Eisenhower's "atoms for peace" plan, Nehru expressed his opposition to participation in an organization dominated by the great powers. (3) 1957—India accepted the International Atomic Energy Agency (IAEA) Statute but argued against a safeguards regime that applied only to importing countries. (4) 1950s—India sought participation and representation in international organizations, because it distrusted great powers' dominated meetings. This distrust went beyond nuclear and disarmament meetings and was expressed in policies concerning Korea and the Indochina conferences as well.[8]

The second approach was shaped by some of Bhabha's successors, who were active in Indian bureaucratic politics and mostly oriented toward the United States, i.e., internationalistic. The outcomes of the state-scientists confrontations depended on the relative strength of the nationalist-internationalist competitive subcultures in India's nuclear history. Generally, when the nationalist subculture prevailed, state-scientists interactions reveal happy convergence; but when the internationalist faction gained the upper hand, state-scientists interactions evolved into ritual confrontations. Both subcultures work within a politically sheltered, secretive, closed politics (i.e., without continuous public scrutiny), and a financially autonomous base. The two allowed successive Indian scientific elites to intervene repeatedly, and often successfully, in policy debates concerning a variety of Indian policies, such as the development of resources (e.g., thorium), nuclear weapons, disarmament, space, energy, and environmental issues. I believe Indian scientific elites have used their scientific knowledge and political skills to advance their respective

7. I have spelled out this idea in A. Kapur, *International Nuclear Proliferation* (New York: Praeger, 1979), 211–13. Also see the discussion in P. Gummett's chapter in J. Simpson, ed., *Nuclear Non-Proliferation: An Agenda for the 1990s* (Cambridge: Cambridge University Press, 1987), esp. 142–44.

8. For these attitudes see S. Bhatia, *India's Nuclear Bomb* (Vikas Publishing House, UP: 1979), 43, 50, 51, 55.

worldviews and conceptions of Indian destiny as well as their personal scientific and institutional interests. They have done so by mobilizing India's political system and social attitudes in support of scientific as well as military, political, and developmental objectives.

The ability to define the national research agenda, to draw on budgetary support for their research activities, to organize state-funded institutes and laboratories, and to participate in international scientific meetings while defining India's position in these meetings, strengthened the domestic political power of these elites. As a consequence, there were no effective mechanisms of political and social control over scientific activities. As to science policy, it reflected the balance of power within the government of India in different historical periods.

Structural Influences

So far, I have addressed personality and ideological factors, such as the close personal relationship between Nehru and Bhabha. Yet structural factors, such as the influence of regional crisis and international developments significantly increased the need for such personalized interactions. Frequent involvement in regional conflicts (the 1962, 1965, and 1971 wars), India's continuous attention to a hostile international environment, and the perception that the great powers were intrusive actors, led to significant levels of investment in dual-use atomic and space technological developments. This dual-use commitment also required extensive international contacts. The nature and level of contacts with the outside world was driven by a search for scientific partners who shared Indian scientific priorities. These contacts were driven by a quest for autonomy, not dependence. The rationale for a strategy to rely on dual use technologies in Indian nuclear and space activities was twofold: (1) Dual-use technology facilitated a plausible deniability of military intent. (2) At the same time it enabled a weapons option should adverse contingencies require such a policy shift in the future. Here the interactions between the Indian state and the international system were triggered by threats; and they facilitated Indian scientific activities through scientific exchanges.

There are two important historical periods in the analysis of structural influences on India's political economy of science.

Pre-1947

The ideology of using state supported science to develop national power emerged during the 1930s and 1940s in a colonial historical context. Three points are important in this period. (1) The British India government had created an infrastructure of industrial and scientific activity in the form of railways, roads, irrigation, post, a telegraph communications network, and a variety of scientific and functional institutes, as noted earlier. This infrastructure was set up by the British mainly to serve their own objectives in India. (2) There was an expectation of an impending change in the nature of the Indian state (from a colony to an independent state) among Indian scientists and politicians and of required changes in the scientific tasks of the new state. (3) There was also an expectation of a fusion between science and national development objectives and massive socioeconomic change to promote the public good. This change required a centrally planned system that would use science to take India into the twentieth century, escape the backwardness of the past, and avoid scientific imperialism and dependency.

The origins of independent Indian science therefore, lay in changing historical state forms and in the political attitudes of India's scientific and political elites. Before 1947, as argued, the function of British military and industrial science and technology was to serve the aims of British imperial domination over India, and not to facilitate Indian economic development or political autonomy. The aim was, in other words, to make India a dependent part of the British global economy. British investment in India's military and industrial research was limited.

Post-1947

After 1947 the state form and its aims changed radically as did the global and national boundaries of Indian science. A new political (attitudinal, organizational, and policy) framework of state-science relations emerged. The shared anticolonial experience among preindependence Indian scientists and political leaders shaped the political consciousness regarding the socioeconomic and political functions of science, and it highlighted the importance of national autonomy. Here a shared political history led to a happy convergence between scientists and Indian political leaders. Both lacked political and budgetary power to implement their ideas in research plans and government policies. Hence the opportunity for ritual confrontations was missing.

The main emphasis of the post-1947 political economy of science was on:

a. The quality of the "Indian mind" and a tradition of devotion to the study of science *and* political statecraft.[9]

b. India's material backwardness and the need to use science for economic development.

9. See D. Chattopadhyaya, ed., *Studies in the History of Science in India,* 2 vols. (New Delhi: Editorial Enterprises, 1982) for Indian scientific activities extending to ancient times. Kautilya's work represents India's realpolitik tradition in ancient India.

c. India's weak international economic and military position and the ambition of Indian political and scientific elites to establish the material base of Indian power in scientific, defense, and foreign affairs. Here state requirements and elites' perceptions inside and outside the government of India (1940s and 1950s) rather than market forces defined the demand to produce and to organize scientific knowledge and its practical applications. The primary producers and consumers of scientific knowledge and its applications were state-sponsored scientific and technical establishments and the government itself.

In the early stages of development of Indian science policy and state-scientists interactions, state interventions in the definition of research priorities was practically nil. In the years before Indian independence "Indian" market forces were linked to British imperial requirements. In the early years after Indian independence Indian market forces were not strong enough to define the demand and production of scientific knowledge. The political consciousness of Indian scientists, and the attitudes of leaders like Nehru and Bose, rather than market forces established the basis of state-scientists interactions.

It is appropriate in general terms to differentiate modern states by the nature of their political and economic organizations as Etel Solingen does in the introductory chapter to this volume. However, these categories are not sufficient to explain the origins and history of Indian state-scientists interactions. In the Indian case, *politicization* has two meanings. First, it meant that political values and political goals drove Indian scientific activities. Second, it meant that scientists used scientific knowledge to engage in "ritual confrontations." Here the Indian scientist played a political role and was not simply working at the cutting edge of science. In addition, because the decision process was secretive and elitist, it was not open to critical public scrutiny. If *politicization* means participation in a public debate, the majority of Indian scientists usually were not political.

The triggers for political mobilization, the membership in the dominant coalition and the results of state-scientists interactions varied during the 1947–80s period. Figure 2 outlines the changing pattern of state-scientists interactions in the nuclear sphere. This outline is not comprehensive; the hard information is buried in secret government files. It should therefore be treated as a suggestive, rather than a definitive framework.

Summing up this section on the relationships between external and internal factors, the "prehistory," is critical to understanding Indian state-scientists interactions. The pre-1947 debates shaped policy and research priorities and produced winners and losers. The big loser was Saha. The big winners were Bhabha, Bhatnagar, and Mahalabonois. Both winners and losers had important international scientific links since the 1930s and 1940s. Thus, it was

Period	Triggers of Action	Dominant Coalition	Results
1947–50s	Changing historic state form; U.S. policies after 1947.	Nehru-Bhabha. The political center is strong and the bureaucracy is strong.	*(a)* Policy of nuclear autonomy and double-track nuclear development. *(b)* Development of technical infrastructure.
1960–66	India-China and India-Pakistan tensions and wars occur.	*(a)* Nehru-Bhabha up to 1964. *(b)* Shastri-Bhabha up to 1966. The political center is weakened by internal and external developments.	Scientists intervene against Nehru's policy injunction against Indian nuclear arms. The intervention is not public. Tension between the policy and the research agenda is apparent.
1966–71	India's internal political situation is volatile, Mrs. Gandhi is engaged in a power struggle, but there are opportunities for bureaucracies to influence India's nuclear and military policies.	*(a)* Mrs. Gandhi and changing set of political and science advisers, depending on issues. The political center is weak, the bureaucracy is divided and politicized, and it is vulnerable to vetoes by foreign-linked (U.S.) internal bureaucratic controversies. (*b*) Foreign (U.S.) constituencies affect, by issue, Mrs. Gandhi-centric dominant coalition activity.	*(a)* "Ritual confrontations" occur on NPT, IAEA, and nuclear supply issues. *(b)* Indian policy oscillates between "autonomy" and "dependence" on NPT/IAEA regime, foreign governments, and nuclear supply issues.
1971–80s	International and regional resistance to development of India's power emerges.	*(a)* Mrs. Gandhi-Sarabhai-Sethna-Ramana. The political center is strong in security affairs. *(b)* Foreign (U.S.) constituencies, as in 1966–71. *(c)* Rajiv Gandhi-Srinivasan, after Mrs. Gandhi's death in 1984. *(d)* Rajiv Gandhi-Iyengar in the late 1980s.	*(a)* Large-scale development is activated by Sarabhai with strong support from Mrs. Gandhi. *(b)* In 1974, India explodes a nuclear device. *(c)* After some oscillation in nuclear and NPT affairs, "autonomy" is restored. As a result of missile developments, the policy prescription against Indian nuclear weapons is undermined significantly.

Fig. 2. State-scientists interactions, 1947–80s

natural for foreign influences to bear on Indian scientific affairs after 1947, with new policy challenges and opportunities. The foreign inputs into Indian scientific affairs were exclusively Western because Indian scientific elites had gained their advanced training in European research organizations. Contacts with Communist countries did not exist. After independence, Nehru's definition of India's nuclear and science policies marked him as one of a few politicians interested in scientific affairs. This was a period, 1947 to 1960, of "happy convergence" between the scientists and the state. The policy boundaries were broad enough to accommodate multiple Indian scientific and political interests. Indian scientists fell into two categories. The "doves" wanted Indian nuclear disarmament and atomic energy work for peaceful uses only. The "hawks" wanted an Indian weapons program and, as an intermediate step, the development of a nuclear weapons option. Doveish scientists justified their worldview by working on peaceful uses of atomic energy, and the hawks expressed themselves by pushing Indian reprocessing activities. But regional, international, and domestic events from 1960 onward put pressure on Nehru and the Indian political system to make difficult choices and to redefine the policy.[10] This process started under Nehru years, but it crystallized after his death in 1966. The wars with China (1962) and Pakistan (1965, 1971), China's emergence as a nuclear power (1964), the development of the NPT (1964–68) and IAEA safeguards regime (1960s), the emergence of a U.S.-USSR detente in NPT affairs (1962), the emergence of bilateral pressures from the United States and Canada on India in the nuclear sphere (1970s), and the growth of clandestine foreign influences in Indian nuclear decision making triggered state-scientists confrontations.

The process and the pattern varied. The period from 1947 through the 1950s was one of political passivity and happy convergence because there was agreement about policy and research priorities. From 1960 onward the process of scientists' intervention in the policy boundaries emerged in the context of external pressures and the growth of Indian nuclear capability. By the mid-1960s India developed a nuclear technical capability and, according to a U.S. government estimate, "The Indians are now in a position to begin nuclear weapons development if they choose to do so."[11]

The scientists' position was led by Bhabha, and targeted against Nehru, the prime minister. Very secretively, Bhabha sought permission to make a nuclear bomb using plutonium. Nehru resisted but kept the option open. The debate was still not settled when Nehru died in 1964.

The 1966–71 period was one of high ambivalence and polarization in state-scientists interactions regarding Indian nuclear affairs. There was an oscillation between a political (public and bureaucratic) tendency to seek autonomy in nuclear and space affairs on the one hand, and a tendency to increase interdependence between a U.S. dominated international security regime and Indian nuclear and space activities on the other. The setting for this intense struggle about the future direction of India's nuclear program lay in a changed internal political context. Indian Congress Party politics were polarized and there was an intense struggle for power between Mrs. Indira Gandhi and her internal political enemies. Mrs. Gandhi's political preoccupa-

10. I have discussed this in my paper, "Nehru's Nuclear Policy" (Nehru Centennial Conference, University of Toronto, October 1989).

11. On Indian nuclear weapons development see *U.S. Department of State,* Director of Intelligence and Research, "Research memorandum, INR-16" (U.S. Department of State, Washington, DC, 14 May 1964).

tions and inexperience in nuclear affairs created opportunities for U.S.-linked Indian scientists and bureaucrats to assert the case for interdependence and Indian denuclearization.

The period from 1971 through the 1980s saw a return to the policy of autonomy in nuclear and space affairs, and an intensified development of space and missile programs following the nuclear test in 1974. These research and policy developments reflected the influence of nationalistic Indian scientists in scientific and political affairs. In the 1930s, before Hiroshima, Indians recognized the importance of the atom. By the 1980s, India possessed nuclear weapons and missile capabilities. Throughout this period happy convergence was followed by ritual confrontations, which, in turn, led to new happy convergence. Despite international pressures under the Bush administration to force India to join the NPT regime, the tilt toward nuclear arms and ballistic missiles appears irreversible.

Research Design Implications

The nationalist orientation shaped by the colonial period and the post-1947 history of interactions between Indian scientific and political leaders requires a modification in Solingen's research design. In her model, the "nature of the state" and the "political economy of science" imply or assume that the two elements are fully developed; the United States seems to be the implicit standard or benchmark. Further it is implied that state interventions define the policy and the research framework of a country's science policy. In India's case, the political thinking of Indian scientists had its origins in British India's colonial context. The first sign of Indian scientists' influence was that the nature of the Indian state and Indian administration after 1947 was based on the priorities of Indian scientists and Indian political leaders prior to 1947. These required heavy state intervention in economic and military planning. The science policy planners were Indian scientists who had been key players in the pre-1947 history and who later became key players in the formulation and management of Indian scientific activities. In both eras the attitude of Indian scientists and Indian leaders was that India missed the first industrial revolution and should not miss the second. The colonial context was decisive in defining the political thinking of Indian scientists as well as the science-oriented thinking of key Indian political leaders. They shared the belief that India needed twentieth-century science and technology to solve current socio-economic problems. In my account there are no Indian or non-Indian institutional buffers, mediators, or umpires between the state and the scientists. The relationship is intimate and continuously so. The political ethos of important Indian government scientists did not conflict with the political goals of India as set out by Indian leaders such as Nehru. Indeed, the political ethos of important Indian political, bureaucratic, and scientific groups was indigenous and nationalistic in its origin. The ritual confrontations concerned attempts by sections of India's scientific and foreign policy bureaucracy, and elements in its political leadership, to rely on foreign solutions for Indian problems, and to dilute the nationalistic role of Indian science in an adverse diplomatic and military environment.

In India's case the "socioeconomic characteristics of the scientific community" are not relevant in defining the state-scientists interactions. Before 1947, the number was small but scientists were internationally recognized. After 1947 this group was divided into "insiders" and "outsiders." Both were in contact with influential scientific elites in India and abroad. However, the "insiders" were Indian scientists who participated in the policy process and who defined the research priorities and strategies of Indian science. This group also had some weight in international scientific and political circles. In contrast, the "outsiders" had weight in scientific circles, and enjoyed limited budgetary support from the Indian government (e.g., Saha). The state's political leaders, bureaucratic administrators, and "insider" scientific administrators controlled the political and budgetary allocations to "outsiders." This distinction, rather than the socioeconomic characteristics of the scientific community, is the relevant consideration in this case. In preindependent British India and independent India, the political and scientific ethos emphasized scientific and political elitism so as to capture and develop pockets of excellence that could aid Indian developmental and security aims. The feeling was, and still is, that it did not really matter whether the scientist was a rich Bhabha or a poor Saha. Both were important, and both types were committed to the development of Indian science; Indian economic, social, and military development; and the making of a new India.

My focus here is on the *intervention strategies* of Indian scientists vis-á-vis the state (government of India), and the development of the *scientist-politician coalitions* that had an authoritative role in developing national plans and Indian science policy. I differ from Solingen in my emphasis on the activities of scientists vis-á-vis a *generally passive state apparatus* in scientific affairs. The rationale for my emphasis is that political elites played a prominent role in the withdrawal of British power from India and in the organization of Indian political power and institutions after independence, but were mostly (with the exception of Nehru) passive in their attitudes to scientific affairs.

Indian Science under Nehru

Nehru and important Indian scientists defined two relationships: (1) between science, technology and economic development; and (2) between peaceful development of Indian atomic energy and nuclear weapons capability. Nehru was the sole decision maker in Indian foreign and scientific affairs up to 1962 because he directed the Foreign Office and held the atomic energy and the planning portfolios. He emphasized the importance of secrecy in atomic affairs, and thus immunized himself and his government against public scrutiny.[12] The scientists-Nehru coalition defined the nature of the state, the policy framework concerning science, the administrative method of scientific policy-making, the budgetary support for scientific activity, the role of science in national planning, and the importance of a mixed economy and the public sector. Nehru's writings reveals his faith in the role of modern science for national development and as a source of Indian military power. He stressed the importance of central planning, and of a strong administrative and political center in the Indian Union taking an all-India view of national interests. Nehru's romantic view of the Bolshevik revolution (in contrast with Nehru's horror regarding Stalinist terror), and his Fabian socialist leanings, led him to stress the need to change backward societies through a planned economy. At the same time, Nehru's policy actions revealed the realpolitik basis of his atomic energy policy. By 1946, the dual-track atomic energy program emphasized the peaceful uses of atomic energy while leaving the military option open. Thus, under Nehru, state-scientists interactions occurred always in the context of two permanent imperatives in Indian policy regarding development and security.

Different scientific fields played important roles in defining the nature of the state and its scientific priorities since 1947. The need of a planning machinery with a scientific basis was advocated and organized largely by Professor P. C. Mahalabonois, a statistician. All planning for the first Indian five year plan took place in Calcutta at the Indian Statistical Institute. Here statisticians played the greater role in defining the normative value of planning and in creating the data base for it. The work of Indian physicists took place in this context.

In India, control in public affairs comes from three different sources: (1) from the right to issue orders through a high administrative post; (2) from an ability to develop an exchange relationship—i.e., by using pressure, compromise, and compensation to accommodate key players' interests. (Here "policy" is the result of such accommodation. In India two kinds of exchange relationships exist: between purely domestic constituencies; and between domestic constituencies who are involved in domestic controversies, and who are simultaneously aligned with foreign constituencies—where the U.S. government is the key external player.) (3) Finally, control is gained from an

12. Nehru publicly emphasized the importance of secrecy in atomic energy on 6 April 1948 when he introduced the Atomic Energy Bill to the Indian Parliament.

ability to persuade, to achieve a meeting of minds via discussion. All three types of control emanate from the state.

Of these methods of control, the first two are most important, but the third is the most democratic. It requires high tolerance of pluralism and consensus. The first approach is the least democratic. The second one, with two variants, is neither democratic nor authoritarian. The first control approach is deeply embedded in India's administrative culture. The government of India works on the imperial principle that the office holder at the top possesses the responsibility and hence he/she possesses the right to command. This principle was also the basis of administration in British India, in Mughal India, and in ancient India. This is the basis of modern Indian administration. This was the formal basis of Nehru's science policy administration, but, as the previous discussion shows, the ritual confrontations in Indian science point to the central importance of the second control method.

Whether the first or the second method was prominent in the history of Indian science decision making after 1947, the administrative structure was such that high level decisions filtered down; lower level debates did not filter upward toward the decision center. The policy debates were among equals in a small decision group at the top. Nehru was not an umpire among competing scientific schools within the government of India. Umpiring was not necessary because the decision-making process was collegial and secretive, and the deviants had been marginalized. The system was not *developed*, if the term implies the existence of publicly competing political and scientific cultures, competing institutional interests, role differentiation, adversarial societal relationships, competing social purposes, mass mobilization, and public confrontations. Such development reveals continuous, organized elite-mass and interelite interactions as well as pluralism, coalition, activity, and in-fighting in scientific and political affairs on a massive societal scale. Instead the scientists-Nehru coalition captured the state apparatus early on. It established its command of the key instruments of state power concerning scientific and foreign affairs. It defined the boundaries of national policy and scientific activity. To the scientists, the state was important because the foremost member of the power elite (Nehru) possessed and provided the political will and the financial resources to support scientific activities. To Nehru the scientists were important because their work enabled his government to refashion India in the image of Nehru's thinking. The coalition was stable in the 1950s because the political and the scientific priorities were compatible and mutually advantageous to the two groups. This discussion reveals the existence of an autonomous coalition, a directorate of Indian scientific affairs in the 1940s and 1950s.

What were the determinants in Indian state-scientists interactions during the Nehru years? I propose a distinction between *precrisis* (the 1930s through

1962) and *postcrisis* (post-1962) relations. The anticipation of India's independence and its upward mobility in the international system produced an anticipation of the need to restructure India's internal economic, institutional, and military strength, and to restructure India's power relations with the outside world by making India an autonomous force in the world. These attitudes led to a high level of political commitment and state support for the development of dual-use science and technology. Regarding secrecy, the system was open. For example, Nehru was willing to rely on foreign scientific advisers (e.g., Professor P. M. S. Blackett). Nuclear science cooperation flourished between scientists in India, Canada, the United States, the United Kingdom and France during the 1950s and the mid-1970s. This indicated a high tolerance of international scientific exchanges. Bhabha was also open in declaring his ambition to take India into the nuclear club. But on the other hand, the system was highly secretive. Nehru was committed to the "built-in defense use if compelled" view of Indian nuclear science, but did not advertise his commitment publicly. The internal Nehru-Bhabha debate in the 1960s about the importance of nuclear arms for India was secret, and it still lies buried in government files. The pattern of Indian technical activities in the nuclear, space, and missile areas is generally known, but the relationships between the technical activities and policy is hidden in secret files and in orally communicated decisions of the political leadership on a need-to-know basis. The method was to not record major decisions in official minutes because of the presumption that Indian government departments were penetrated by spies.

Despite the secrecy, the autonomy of the dominant state-scientists coalition of the Nehru era was fractured as a result of increasing political pressures in Indian politics, and the rise of bureaucratic politics since the 1960s. External developments effected these changes. Figure 2 shows the circumstances in which the scientist-Nehru coalition started to lose its closed, secretive, and totally autonomous character. External developments (the India-China war, the India-Pakistan war, China's nuclear testing, and the NPT) put pressure on Indian policy positions and rhetoric, and as a result a public debate followed. These pressures became entrenched in India's political system, and they were irreversible. The present position is that India's policies since the Nehru years have changed. For Nehru, nuclear disarmament was an urgent question; for his successors it is a distant goal. For Nehru the public commitment against Indian nuclear arms, and for peaceful uses of atomic energy, was firm; for his successors the move is on to acquire nuclear weapons and missile capability. This position is the result of the ritual confrontations of the 1960s, 1970s, and 1980s that started in the last phase of the Nehru era, although the debates never actually took shape until after Nehru's death in 1964.

Post-Nehru Indian Science

The ethos of the scientific, political, security, economic, and nongovernmental (public, academic, and press) constituencies started to diverge. The autonomy of the scientist-politician coalitions came under attack. Public debates reflected and reinforced internal bureaucratic debates about India's nuclear and disarmament policies that involved Bhabha, Nehru, and the civil servants. Two new elements shaped these confrontations. First, following the India-China war of 1962, and China's nuclear test in 1964, there was an increased public, parliamentary, nongovernmental opposition to the Nehru posture that stressed a primary reliance on peaceful uses of nuclear energy. Out of this public opposition came additional pressure to move toward the development of an Indian nuclear weapons program. Second, by the mid-1960s, the U.S. government was actively involved in internal bureaucratic battles in India, pushing aggressively for Indian denuclearization. (Declassified U.S. government cable traffic shows heightened U.S. awareness of India's nuclear ambitions at this time.) Two lines of nuclear thinking emerged in India at the time. The U.S.-linked bureaucratic coalition in India was in the Indian Ministry of External Affairs and in the Indian Atomic Energy Department and favored the elimination of Indian nuclear autonomy, as well as a yielding to the international nonproliferation regime. The second line sought an Indian bomb or, at least, maintainance of India's nuclear weapons option. The latter approach was expressed by Indian diplomatic and strategic planners who opposed the pro-U.S. approach.

Between the early 1960s and the early 1970s, Indian nuclear decision making and international conference diplomacy oscillated between these two policy lines and social tendencies. Neither side won the debate completely; both lines coexisted inside the Indian government and in public debate, but the bomb lobby appeared to maintain an edge in the decision process. The controversy led to an intensified growth in dual-use technologies. Out of the 1962–71 controversies came Vikram Sarabhai's space program, and the preparations for a nuclear test that occurred in 1974. Here military applications of Indian nuclear science were investigated secretly, and the costs of a weapons program were checked. The pattern here is that internal controversy and external pressure to join the NPT regime led to an intensified development under the "dual-use" cover of the military content of Indian nuclear and space activities. This pattern continued in the 1980s when the decision to develop missiles led to the acquisition of intermediate range missile technology.

In sum, as a result of external pressures on India and internal controversy about nuclear policy, three patterns of behavior are revealed. First, Indian scientists have been politically active in scientific and policy spheres, but, after 1947, the nature of the activism depended on whether or not the scientist had foreign links. Second, Indian policies have oscillated as a result of state-scientists interactions. For example, Prime Minister Shastri moved toward a bomb test in 1965, but Mrs. Gandhi's first draft in 1967 favored the NPT, and she backed away from signing the treaty in view of internal bureaucratic and Cabinet opposition. Mrs. Gandhi opposed Indian nuclear testing in 1966 but approved it in the early 1970s. Mrs. Gandhi opposed a weapons program but she supported the development of an Indian space infrastructure and weapons-related research and development. Prime Minister Rajiv Gandhi wanted India to join the NPT but eventually accepted scientific and military advice to promote Indian missile development. Finally, despite the oscillations, the pattern is still one of incremental growth of the military dimension of Indian nuclear and space activities, even as international pressures to denuclearize India have increased in the post-Cold War era.

APPENDIX A

History of Scientific Activity in India before 1947

1876	The Indian Association for the Cultivation of Science (Calcutta)
1890	Botanical Survey of India
1899	Haffkine Institute
1905	Agriculture Research Institute (Pusa, Bihar; transferred to Delhi, 1936)
1906	Forest Research Institute, (Dehra Dun)
	Central Research Institute for Medicine
	Medical Research (Kasauli)
1911	Tea Research Institute
	Indian Institute of Science (Bangalore, supported by J. N. Tata)
1913	Indian Science Congress Association
1916	Zoological Survey of India
1917	Bose Institute (Calcutta)
1921	School of Tropical Medicine (Calcutta)
1923	Dairy Research Institute (Bangalore)
1924	Cotton Technology Lab (Bombay)
1927	Malaria Institute (Delhi)
1928	Nutrition Research Institute (Conoor)
1930	Academy of Sciences (Allahabad)
1931	Indian Statistical Institute

Source: A. Parathasarthi and Baldev Singh, "Science in India," paper presented at the "India the First Ten Years" conference, University of Texas, Austin, 6–9 December, 1990.

1934	Institute of Public Health and Hygiene (Calcutta) Indian Academy of Sciences (Bangalore)
1935	National Institute of Sciences
1936	Institute of Sugar Technology (Kanpur)
1938	National Planning Committee (India Congress Party)
1939	Jute Research Institute (Calcutta)
1942	Council of Industrial and Scientific Research (Government of India)
1943	Professor A. V. Hill, review of postwar Indian scientific and industrial research, at the invitation of the Indian government
1945	The Tata Institute of Fundamental Research (Bombay, supported by D. Tata Trust) Education Department review to consider establishment of Institutes of Technology on MIT lines.
1946	Research Committee on Atomic Energy (Government of India) Institute of Paleobotony (Lucknow)
1947	Shri Ram Institute for General Industrial Research (Delhi)

The Political Economy of Science and Technology in Israel: Mutual Interests and Common Perspectives

Gerald M. Steinberg

The Centrality of Science and Technology in Israel

Although Israel is a small country, both demographically and economically, the emphasis on science and technology is very pronounced. Scientific and technological activities are central to the economy, political system, and social structure. There are over sixteen thousand scientists and engineers in Israel, and sixty-nine per ten thousand in the labor force are engaged in R&D related activities.[1] On a per capita basis, these totals are among the highest in the world. In 1990, approximately 4.5 percent of GNP went to R&D, compared to 2.7 percent in the United States, 2.3 percent in Britain, and 2.6 percent in Japan.[2] Members of the scientific community, such as Moshe Arens and Yuval Neeman have also become major political figures, and in many ways, this community is closely linked to the political and social elite.

Science and technology also play important ideological roles in Israel. Forty years ago, the economy was largely based on agriculture. Science and technology were seen as the major path to industrialization and modernization. Ideologically, scientific development was perceived as a basic component of modern political Zionism. In his visionary tract, *Altneuland* (Old-New land), the founder of political Zionism, Theodore Herzl, emphasized the role of science and technology in the revival of the Jewish people.[3] Similarly,

---

1. E. Tal and E. Katzir, "The Case of Israel," in *Worldwide Science and Technology Advice to the Highest Levels of Government,* ed. William Golden (Oxford: Pergamon Press, 1990).

2. The high level of R&D in the period 1988–90 reflected the major investment in the Lavi combat aircraft project. Prior to this period, R&D consumed approximately 3.5 percent of the GNP. See A. Keynan, *Science and Israel's Future: A Blueprint for Revitalizing Basic Research and Strengthening Science Based Industry* (Jerusalem: Jerusalem Institute for Israel Studies, 1988), 22.

3. T. Herzl, *Altneuland* (Old new land) (Reprint ed. Jerusalem: M. Newman, 1960); see also Baruch Kimmerling, *Zionism and Economy* (Cambridge, MA: Schenkman, 1982).

David Ben-Gurion, the leader of the independence movement and Israel's first prime minister, frequently spoke and wrote of the centrality of scientific development in the rebirth of the nation. Zionist leaders also viewed scientific and technological achievement as a source of international political power and leverage. Many attributed the 1917 Balfour Declaration, in which the British government recognized Jewish aspirations for a "national home," to the scientific contributions of Zionist leader Chaim Weizmann to the war effort.[4]

This ideological and political emphasis provided the impetus for the early creation of advanced research institutes, such as the Weizmann Institute and the Israel Institute of Technology (Technion), and of the system of universities in the country. These institutions have received significant support, and the Israeli leadership continues to view research and development as central factors in the economic, social, and political development of the country.[5] Political leaders view scientific and technological development—as expressed in projects such as the Lavi combat aircraft program, the Shavit launcher and Ofek satellite programs, and the Arrow antitactical ballistic missile (ATBM)—as a source of national pride and independence.

Israeli foreign policy and national security are also closely linked to the activities of the scientific and technological community. In these policy arenas, the importance of science and technology in Israel is even greater than the role of these factors in the United States in the 1950s and 1960s.[6] The Israeli military has used qualitative superiority, in the form of advanced military technology and skilled personnel, to offset the massive demographic and financial advantages of the Arab states. These advantages have led the Ministry of Defense to invest heavily in military-related R&D, and in the development of the necessary infrastructure. A large proportion of the nation's

---

4. In most accounts of the Balfour Declaration, the role of Weizmann's achievements as a chemist are exaggerated. This exaggeration is, in itself, an indication of the perception that scientists, by virtue of their achievements, are able to wield power in the international political system. For a detailed analysis of the perception of Weizmann's role in the formulation of the Balfour Declaration, see David Vital, *Zionism: The Crucial Phase* (Oxford: Oxford University Press, 1987).

5. The perceived importance of scientific and technological activity in economic and social development is demonstrated in government policy. See, for example, Council for National Planning, *International Developments and Their Impact on Israel and the Middle East in the Next Decade* (Jerusalem: Ministry of Economy and Planning, 1988); Council for National Planning, *Technological Culture as a Condition for Economic Growth* (Jerusalem: Ministry of Economy and Planning, 1988); and J. Jortner, *Scientific Research in Israel; Perils and Challenges* (Jerusalem: Israel Academy of Sciences and Humanities, 1988).

6. See Eugene Skolnikoff, "Science, Technology, and the International System," in *Science, Technology, and Society: A Cross Disciplinary Perspective,* ed. Ina Speigel-Rosing and Derek John De Solla Price (Beverly Hills: Sage, 1977).

scientists are linked in some way to the defense establishment and armed forces.

In many ways, the views of the members of the Israeli scientific community on science and technology-related social and political issues are relatively homogenous. There are few debates or controversies among the scientists, even on vital issues such as nuclear deterrence or the environment. This high level of homogeneity is particularly noteworthy given the generally conflictual nature of Israeli society and political debate. It may also be a reflection of the degree to which Israeli scientists still hold the classical Baconian view that science is apolitical and "above" politics.[7]

Indeed, despite the general coincidence of political and social views between the scientific community and the governing elite, Israeli scientists are often accused of pursuing professional goals that are irrelevant to national needs. Israel is a mobilized society in which resources are focused on national goals such as defense, immigrant absorption, and economic development. In contrast, in a broad sense, the scientific community tends to be independent and universalist, based on internally selected goals and structures, such as peer review. Israeli scientists, like their counterparts in the United States and other societies, view the state as a source of resources alone.[8] Outside of the government-controlled defense sector, state intervention in attempting to define scientific goals and research projects is seen as interventionist.

To a large degree, the continuing strength of universalist values is a reflection of the role that prominent Jewish scientists played in promoting these values. In the last century, Jewish scientists in Europe and the Soviet Union were alienated and victimized by nationalism. In Nazi Germany, the teaching of "Jewish physics" (relativity) was anathema, and Jewish scientists were forced to flee. The emphasis on universalism of science can be seen, to some degree, as a response to this historic experience. Similarly, the scientific community invoked the values of the "universalism of science" in demanding freedom for Jewish scientists in the Soviet Union who had been isolated for decades under Stalin and his successors. Many of these scientists moved to Israel, and the universalism of science remains an important element in the Israeli scientific community.

---

7. See Sanford Lakoff, "Scientists, Technologists, and Political Power," in Speigel-Rosing and De Solla Price, (see n. 6); and R. C. Wood, "Science and Politics: The Rise of an Apolitical Elite," in *Scientists and National Policy Making,* ed. R. Gilpin and C. Wright (New York: Columbia University Press, 1964).

8. Scientists and analysts of R&D in Israel have generally ignored the relationship between the Israeli political and social structure and the scientific community. See, for example, E. Tal and Y. Ezrachi, *Science Policy and Development: The Case of Israel* (Jerusalem: National Council for Research and Development, 1972); Tal and Katzir, "The Case of Israel" (see n. 1); and Keynan, *Science and Israel's Future* (see n. 2).

Science and the State: The Political-Economic Perspective

The Structure of the Israeli Political Economy

Israel has existed as a modern state for less than five decades, and the central political, military, and economic conditions and institutions have been transformed significantly since 1948. Given the centrality of science and technology, the evolution of political and economic conditions and institutions have necessarily been accompanied by, and, in many cases, directly involved in, changes in the role of science, and in the interaction between science and the state.

For the first thirty years of its existence (and for the prestate decades of the New Yishuv), Israel was highly centralized, both politically and economically. The country was dominated by an all-encompassing elite that consisted of the socialist (or social-democratic) Labor party (Mapai, at times in a coalition with smaller allied parties such as Mapam and Ahdut Avodah, or Rafi), and the Histadrut General Labor Federation. This elite controlled educational institutions, agricultural marketing cooperatives, and a large part of the industrial base. Thus, although formally and juridically pluralist, Israeli society and the economy were, in fact, largely closed and monopolistic.

At the same time, the economy was never characterized by central planning or resource allocation on a scale similar to the Soviet Union or to developing states such as India and Brazil. As in Italy and France in the 1950s and 1960s, Israel's government and dominant labor federation (the Histadrut), in combination and often indistinguishable, dominated critical sectors of the economy (energy, transportation, etc.) As will be shown subsequently, these structural conditions were reflected in economic policy and the application of technology to development.

In the 1970s, following the "earthquake" of the 1973 Yom Kippur War, which greatly weakened the elite structure, political competition increased and the pluralistic environment was strengthened. In the 1977 elections, the Labor alignment lost control of the government, and through the 1980s, was only able to regain some power as a partner in a broad-based coalition with the Likud and smaller parties. The Likud leadership, in coalition with the Israeli Liberal party, took steps to reduce the role of the state in the economy, and allowed the role of the market to increase. The central role of the Histadrut in the economy has also diminished significantly. Thus, in the past decade, Israel has begun the transformation from an operationally monopolistic centrally planned system to a mixed, or pluralist market-oriented system.

Economics, Science, and the State

In the first chapter of this volume, Etel Solingen notes that the political economy of science is "the set of interacting local and international, state, and market processes affecting the demand and production of scientific knowledge, its location, and the associated distribution of costs and benefits." In the Israeli case, the perceived importance of science and technology in economic and military development has given the scientific community ready access to political and economic leaders. This situation also provided preferred access to resources, as well as extensive control over the distribution of these resources. As noted, Ben-Gurion sought to apply science and technology to the solution of many of Israel's economic and security problems. He created the Science Corps in 1948, at the beginning of Israel's War of Independence, and provided the Corps with a relatively high level of scarce national resources. The National Council for Research and Development Council (NCRD) was located in the Office of the Prime Minister. (In 1982, the NCRD was transferred into the newly created Ministry for Science and Technology.)

As in the case of many other industrialized and industrializing countries, the scientific community and the state are highly interdependent, and the degree of interdependence in the Israeli case is particularly pronounced. A very high proportion of R&D activity is supported by the state. Over 60 percent of national R&D funds are provided by the government, through institutions such as the Office of the Chief Scientist of the Ministry of Trade and Industry, the Israel Investment Center, and the Ministry of Defense. Some 22 percent of R&D funds are provided by industry, compared to an average of about 50 percent in other developed states. Industry employs one-quarter of the R&D manpower, as compared with 72 percent in the United States and 62 percent in Japan.[9] (Some 50 percent of national R&D funds go to the military sector, approximately the same level as in the United States, compared to 22 percent for China, and less than 5 percent for Japan.)[10]

Within the industrial sector, most of the R&D is performed by government-owned firms, such as the Israel Aircraft Industries (IAI), which is controlled by the Ministry of Defense and is the country's largest industrial employer, the National Armaments Development Authority (Rafael), and Israel Chemicals. Other major technology-intensive firms, such as Tadiran (electronics) are owned by the Histadrut, and are also closely linked to the government.

Universities perform 30 percent of national R&D, which is relatively

9. Tal and Katzir, "The Case of Israel" (see n. 1).

10. Keynan, *Science and Israel's Future* (see n. 2), 22.

high (compared to 12 percent in the United States, for example) but less than previous levels. In the 1960s, universities accounted for 65 percent of civil research.[11] In general, while the universities receive the bulk of their support from the state, they are largely free to set research agendas and priorities. The university research system is closely linked to the state Council for Higher Education and the University Planning and Grants Committee (which is modeled on a similar British system). These institutions, as well as the university structure and the Academy of Scientists, cushion the individual scientists and researchers from external interference and provide for the distribution of resources on the basis of peer review. In 1989, the Ministry of Education sought to become involved in the process of university allocations, but this effort was successfully rejected by the universities. Thus, despite the very high level of dependence on the state for resources, the Israeli scientific community is still relatively autonomous, at least at the level of basic research.

Research and Development and Economic Policy

The economic and social development of the state, and the process of modernization in the Zionist movement and in Israel, is largely based on and linked to science and technology.[12] The state has provided significant resources to the scientific community in order to stimulate economic development, and, as noted previously, over 60 percent of R&D (including some of the university-based research) is funded through the national budget. (All universities in Israel are state institutions and are funded through state allocations, donations, and student tuition fees.) Science and technology-based industries are a major aspect of the strategy for industrialization, and for the development of exports. In many cases, the strategy for creating new employment opportunities is synonymous with the development of technologically advanced industries.

In the past decade, the Office of the Chief Scientist (OCS) and the Israel Investment Center have emerged as the primary vehicles for government support of industrial R&D. Together, these agencies provide hundreds of millions of dollars in direct subsidies for R&D projects to firms. The OCS provides from one-half to two-thirds of the funds for "approved projects," while the firms put up matching funds for the rest. Although these funds are

11. Ibid., 16.

12. See, for example, David Ben-Gurion, *B'Maarakhah* (Tel Aviv: Am Oved, 1957), and *Zikhronot* (Tel Aviv: Am Oved, 1971–73). For empirical evidence of the role of technology in Israeli economic development, see Tuvia Blumenthal, "R&D in Israeli Industry," *Research Policy* 7 (1978): 62–87.

ostensibly designed to promote specific high-risk R&D ventures and projects by small firms with little capital for such undertakings, in fact, up to 60 percent of the allocations go to large, well-established government and Histadrut firms, such as IAI, Tadiran, Israel Chemicals, etc., every year, regardless of the nature of the projects involved. In other words, the system of support for applied industrial research has become highly bureaucratized and institutionalized, with the type of patron-client relations that are typical of many such government subsidy programs. The government agencies have close professional, political, and social relationships with the technical personnel and management of the recipient firms. Allocations are divided informally, based largely on previous divisions, and new firms have great difficulty "breaking in."

Similarly, the Israel Investment Center (IIC), which is linked to the Ministry of Commerce and Industry, like the OCS, also provides direct subsidies to firms. The IIC's mandate is to broadly develop industry and employment in the country. In practice, this agency has also focused on technology-intensive industries. While some of the IIC funds go to subsidize foreign investment in Israel (firms such as Intel and National Semiconductor), many of the same firms that are funded by the OCS, such as IAI, are also subsidized by the IIC. The patron-client relationship between the IIC and the recipient firms is similar to that developed in the case of the OCS.

The extensive military R&D is also designed to promote economic as well as national security objectives. Over one-third of industrial employment in Israel is centered in the military sector, and technology-based military products account for approximately $1.5 billion in exports annually, or about 12 percent of total exports (20 percent excluding diamonds and agricultural products).[13] Military technology has also been seen as "the leading edge" of technology-based export industries, which account for some $2.5 to $3 billion in total exports annually (including chemicals, electronics, and the military sector).

Science and technology are also considered crucial in the area of immigrant absorption, on the one hand, and emigration, on the other. Israel produces many more scientists and engineers than can be employed by the universities and industry, and, as a result, many members of this community emigrate in order to seek employment in their professions. This might be considered to be sign of overinvestment in advanced education and training in

---

13. Estimated on the basis of data from the *Statistical Abstract of Israel: 1989* (Jerusalem: Central Bureau of Statistics, 1989), 249–50; and *Annual Report, Bank of Israel: 1991,* (Jerusalem: Bank of Israel, 1992), 330. Official statistics regarding defense and military exports are widely believed to be less than the actual sales and transfers, and many reports of major multibillion transactions with China and other states have been published.

the natural sciences and medicine, but the cultural and social value placed on scientific professions expands consideration of this issue beyond the narrow economic issues.

Instead, politicians and members of the scientific establishment call for government policies designed to create jobs in order to end the "brain drain." Furthermore, from the perspective of Zionist ideology, emigration from Israel is an ideological affront, and it is considered the duty of the government to provide resources in order to halt this process. The Ministry of Education and the Ministry of Science and Development have allocated funds specifically for the purpose of returning expatriate scientists.

The Israeli government has also made major efforts to provide jobs and other resources for immigrants. Between 1989 and 1991, over four hundred thousand immigrants arrived, increasing the Israeli population by 10 percent. (By the mid-1990s, at least one million Jews are expected to have immigrated from Russia and the other republics that constituted the Soviet Union.) A very high percentage of this group have scientific or technical training. Special funds have been created to allow the universities to create new positions and research programs for these immigrants. Although all immigrants (*olim*) are given special assistance, the additional support for scientists and engineers indicates the particular political and ideological importance attached to the absorption of this group.

Thus, in many ways, the Israeli government and economic structure have become highly dependent on R&D and on the scientific and technological community for the creation of jobs, economic development, exports, and immigrant absorption. Through the emphasis on R&D, the economy has grown and developed relatively rapidly and successfully. As a result, the political and social leadership, including most parties and major figures, continue to view science and technology as the key to national development and growth. This perception is expressed in terms of support for highly ambitious projects such as the Lavi combat aircraft, the Mediterranean–Dead Sea Canal, the Arrow ATBM missile, and the undifferentiated subsidies provided for industrial R&D discussed previously.

Policy for Science and Technology

Analysis of the political economy of science also includes the interaction of politics, economics, and the role of the scientific community in establishing policy for science.[14] Issues include the processes by which resources for R&D are distributed among the competing claimants, diversification, social and geographical concentration, and leadership patterns.

14. See Solingen in this volume.

In the Israeli case, many of the major institutions for R&D were established before the state was founded in 1948. They were largely autonomous; and science and technology policy in the 1950s has been described as "a period of largely spontaneous and uncontrolled growth, characterized on the one hand by initiative on the part of bodies within and outside government ministries, and on the other hand by the absence of overall planning and coordination."[15]

In 1949, shortly after the establishment of the state, the government created the Scientific Council to serve as a central advisory body. The Council had little impact on established processes, but it did create new research institutes in applied physics, fiber research, geology, and arid-zone research. In general, however, Israeli R&D in this period was independent of the government, and heavily oriented toward basic and theoretical research. Critics charged the scientific community and the government with neglecting applied research that could alleviate the national problems caused by massive immigration (the ingathering of the exiles) and the constant military threat.

In response, in 1959 the government created the National Council for Research and Development (NCRD), which was charged with developing a national R&D policy, coordinating activities, and administering the institutes created earlier. This agency was located in the Office of the Prime Minister, in order to insure access and resources. In practice, however, the NCRD, like its predecessor, the Scientific Council, had little independent power or budgetary authority, so that it was unable to influence the universities or government ministries. Civil R&D in Israel continued to emphasize basic research and was largely decentralized throughout this period.

Continued demands for redirection of the relatively large scientific and technological effort toward national needs led to yet another reorganization effort in the late 1960s. In this round, the individual government ministries were recognized as the core of power in the state, and as a result, positions of chief scientists were created in most major ministries, including the Ministries of Commerce and Industry, Education, Defense, Agriculture, and so on. Chief scientists were expected to lobby for increased funds for applied R&D within their ministries, and to control and supervise research programs.[16] Some, such as the Office of Chief Scientist in the Ministry of Commerce and Industry, were relatively successful, largely as a result of their connections to the political system. For example, the growth of R&D funds in this case was linked to the fact that Haim Bar-Lev, who had served as Chief-of-Staff of the

15. Tal and Katzir, "The Case of Israel" (see n. 1), 5.

16. Ibid.; see also Committee to Inquire into the Organization and Administration of Governmental Research, *Report of the Committee*, E. Katchalski (Katzir), chairman, (Jerusalem, 1968).

Israeli Defense Forces, became Minister of Commerce and Industry. He brought the former head of military R&D, Yitzchak Yaakov, with him to be chief scientist in the ministry. Both belonged to Israel's political elite, thus insuring a high level of budgetary allocations.

The system of chief scientists succeeded in greatly increasing interest in and support for applied research among the various ministries and industries. At the same time, the NCRD was expected to play a coordinating role, but, given the political power and independence of the individual ministers and ministries, this was impractical. By the mid-1970s, the chief scientists of the major ministries were the primary determinants of applied R&D policy in Israel.

During this period, basic research has remained under the control of the universities and of the Israeli Academy of Sciences. The academy is an independent body, but has taken on the task of coordinating basic research, and lobbying the government for increased funds in this area.

In 1982, Yuval Neeman, one of Israel's preeminent physicists, was elected to the Knesset as head of the Tehiya party and became a cabinet minister. To accommodate him in a broad coalition government, the Ministry of Science and Development was created. The NCRD was incorporated into this ministry, an incorporation that provided the agency with greater visibility and improved its ability to coordinate activities. In practice, however, the Ministry of Science and Development has had little impact on the structure of science and technology in Israel. Reflecting the acceptance of the universalist view of the scientific community, basic R&D has remained decentralized and under the control of the individual universities and researchers (for basic R&D), and applied research continues to be determined by the individual ministries.

The Scientific, Technological, and Political Elite

Many analysts have studied the links between knowledge and power in modern industrial societies. These studies have shown that in many cases, scientific and technological "experts" have been able to turn their professional knowledge into political power.[17]

From the beginning, members of the Israeli scientific community were part of the social and political elite, and despite fundamental political and economic changes, their position has continued without interruption. The respect and importance given to academics and intellectuals in Jewish society also continues, and a position on the faculty of a major university was and

17. See Sanford A. Lakoff, ed., *Knowledge and Power: Essays on Science and Government* (New York: Free Press, 1966).

continues to be highly esteemed. (This is in spite of a strain of anti-intellectualism that accompanied Labor Zionism's rejection of the traditional diaspora society, with its emphasis on "unproductive" scholarship, and rejection of "worldly materialism.") Dr. Chaim Weizmann, the head of the Zionist movement for many years and Israel's first president, was an accomplished chemist.[18] After Weizmann's death, the position of president was offered to Albert Einstein. Dr. Ephraim Katzir, a prominent biologist, also served as president and was selected as a result of his scientific achievements. The office of president is largely ceremonial, but the selection of scientists for such a position is an indication of the importance of science in Israeli society, the esteem in which scientists are held, and the close links between the "political class" and the scientific community.

At the same time, at lower policy-making levels, scientific knowledge, in and of itself, does not provide a direct link to political power. The state and quasi-state institutions, such as the Histadrut Labor Federation and the Jewish Agency, dominate most aspects of the social system and power structure. These institutions, in turn, are controlled by political parties. Individual scientists and engineers have access to power within the structure of the party system and as part of the ruling elite, but not by virtue of their knowledge or scientific achievements. (A detailed study of the role of physicians shows that this professional group of "experts" has almost no impact on health policy in Israel.)[19]

In Israel, the type of institutionalized framework for scientific and technological advice that is found in many advanced industrialized states is also very limited. As noted, there is some institutionalization at the ministerial level, and the NCRD plays a minimal role in coordination. There is, however, no structural equivalent to the U.S. White House Science Council or the U.S. Office of Technology Assessment, for example, to provide advise and evaluation of scientific and technological policy at the highest level.[20]

Instead, high-level interaction between the top decision makers and the scientific community in Israel is based largely on informal personal connections among the elite. As noted, well-known scientists are included within the

18. Weizmann was chosen as Israel's first president primarily in recognition of his role as the leader of the Zionist movement for many years. In this sense, his scientific achievements were secondary.

19. Yael Yishai, *Kocha Shel Mumchiut: HaHistadrut Harefuit Biyisrael* (The power of expertise: The Israeli medical association) (Jerusalem: Jerusalem Institute for Israel Studies, 1990).

20. For an analysis of the development of the scientific advisory structure in the United States, see Robert Gilpin and Christopher Wright, *Scientists and National Policy-Making* (New York: Columbia University Press, 1964); and William T. Golden, ed., *Science and Technology Advice to the President, Congress, and Judiciary* (New York: Pergamon, 1988).

boundaries of the social and political elite, and the economic and military importance of these factors highlight the potential political role of the scientific and technological community. Ben-Gurion had close personal relations with many of the nation's leading scientists and met with them frequently. Since 1977, the personal links between the political leadership and the scientific community have been more impersonal, but the broader relationship between these two groups has been maintained.

In the past decade, individual members of the scientific community, particularly those connected with military R&D such as Neeman and Arens, have become part of the political leadership, and assumed independent political roles. Neeman studied engineering at the Technion and later became a theoretical particle physicist. He worked with Murray Gelman at the California Institute of Technology and made important contributions to the application of group theory in elementary particle physics. He was also closely involved with the Israeli Ministry of Defense, served as Deputy Director of Intelligence and Military Attache in London, and became head of Tel Aviv University. At the same time, Neeman was active in Israeli politics, was a founder of the Ha'Teyiha (Renaissance) party, was elected to the Knesset in 1981, and served in the cabinet as Minister of Science and Development and Minister of Energy.

Arens studied aeronautical engineering at MIT, returned to teach at the Technion, became a manager in the aeronautical engineering division at Israel Aircraft Industries (IAI) and headed the Kfir combat fighter production effort. Like Neeman, he became active in politics, was elected to the Knesset in the Likud party, and has served as Ambassador to the United States, Foreign Minister, and Minister of Defense.

Neeman and Arens were able to rise relatively rapidly in the political hierarchy in part because of their scientific and technological achievements, which automatically made them a part of the Israeli decision-making elite. Both were connected with the defense establishment, which provided them with greater political visibility and access. However, both exercised power by virtue of their political positions, rather than their expertise.

It is also important to note that both Arens and Neeman have combined their political and technological interests, and the boundary between these functions remains unclear. As a manager in IAI, Arens pursued the development of Israeli combat aircraft, and, when he became Ambassador to the U.S. and Minister of Defense, he continued to support the Lavi effort, even when the costs of this project made it untenable. (When the U.S. Department of Defense sought to limit U.S. support for the Lavi, Arens intervened with the U.S. Congress. As domestic criticism of the costs of the Lavi mounted, Arens argued that the technological performance of this aircraft would ultimately lead the U.S. military to acquire the Lavi. When the project was cancelled in

1987, Arens resigned but returned to the Cabinet in 1988 as Foreign Minister.)[21]

Similarly, Neeman has been active in promoting science and technology related projects. He was a major supporter of the Mediterranean–Dead Sea Canal Project (which was ended with little progress in 1985), the Lavi, and, more recently, Israel's space efforts.

As prominent members of the scientific and technological community, Neeman and Arens, as well as others in similar positions, are generally considered experts in issues relating to scientific or technological policy. Many cabinet members relied on the evaluations of Neeman and Arens with regard to the Lavi and the Canal projects, without acknowledging that close links with the scientific and technological community—and a tendency for "technological enthusiasm" (see subsequent discussion)—might lead to distortions in their evaluations. Given the homogeneous nature of the scientific community, in which conflict is generally absent, very few individuals challenged these views.

The Impact of Israel's International Position on Science and the Scientific Community

Security requirements, and the constant threat of warfare in the region, have strongly influenced the direction of Israeli R&D and science and technology. As noted, Israeli strategy emphasizes technological development in order to maintain a qualitative advantage over the Arab forces, thereby attempting to offset large quantitative and demographic disadvantages. This situation contributed to the initial development of the defense industries and their subsequent growth.

Over one-half of national R&D expenditures go toward military related programs, which is very large by any standard. This emphasis on military applications is reflected in the fact that 78 percent of all the national R&D expenditure is focused on the electrical, electronics, and aerospace sectors. (In Israel, both sectors include major military components and are based on foundations created for the military, although civil R&D in both sectors has increased in recent years.)

The Ministry of Defense and Israel Defense Forces (IDF) have consistently supported the development of local technological add-ons and upgrades (electronics and avionics) to improve the capability of imported weapons platforms, to adapt them to the Israeli requirements and environment, and to extend the lifetime of these weapons. The Science Corps and its successors

21. Gerald M. Steinberg, "Large Scale National Projects as Political Symbols: The Case of Israel," *Comparative Politics* 19, no. 3 (April 1987): 331–46.

have also developed specific weapons systems and related technology not available elsewhere, such as mini-RPVs to provide real-time images and data to battlefield commanders, advanced decoys, ship-to-ship missiles, battlefield communications, and so on. When the major suppliers of main battle tanks (Britain in particular) refused to provide such weapons to Israel in the late 1960s, Israel produced the Merkava main battle tank, including advanced computerized fire control and other electronics systems. In the past few years, IAI and Rafael have been developing antitactical ballistic missile technology (the Arrow and AB-10). Technologically and militarily, these innovations have been very successful.

The foundation for Israel's nuclear capability was created in the 1950s under Ben-Gurion, who, like many other political leaders of the time, saw nuclear energy as the key to both economic development and national security. In the case of Israel, in particular, a nuclear weapons capability was and still is perceived by many military and political decision makers as the "ultimate guarantee" of national survival in the face of the overwhelming Arab demographic and economic advantage.[22] The government has supported the growth of the Israeli nuclear program, including the reactor complex at Dimona in order to advance specific political and military objectives.

As a result, in Israel, as in other states with a heavy emphasis on military R&D, such as the United States, government support for this sector, and related industries and fields, has skewed the national R&D system. Beginning in the late 1960s, support for military aircraft development led to a greatly increased demand in this sector. When IAI, which is Israel's largest industrial employer, began to cut back its workforce in the late 1960s, the Israeli government supported the development of the Arava STOL aircraft primarily in order to prevent the loss of these workers and the infrastructure in this area. Given the very limited labor market in Israel, many skilled engineers would have left the country. The Lavi project was motivated, in part, by similar concerns.

While the government initially supported relatively large-scale investment in civil nuclear research, applications, and facilities, economic limitations have led to subsequent reductions in these programs. (The facilities and personnel at the nuclear reactor complex at Nahal Sorek have been put on a commercial basis, and units that are not profitable have been reduced or even closed.) In 1987, the government ended the Lavi combat aircraft project due to a lack of funds for production, which led to an immediate reduction of 25 percent of IAI's workforce. Although some of the "slack" has been taken up in the Arrow antitactical ballistic missile program, this is much smaller in

22. For a detailed discussion of the Israeli nuclear program, see Shai Feldman, *Israeli Nuclear Deterrence* (New York: Columbia University Press, 1982).

scope. Government "sheltering" of aeronautical engineers was greatly reduced as the cost became too great.

At the same time, the large military-related R&D sector is largely separate from basic and civil R&D. Although the state is the main source of funding for basic research, as noted previously, agendas and priorities are largely determined independently by the Israeli Academy of Sciences (through peer group evaluation of research proposals) and the university research system. Similarly, civil research priorities are set by industry (see previously) and by the chief scientists in many government ministries. In contrast, the agenda of the much larger military R&D sector is determined by military contracts and requirements.

"Openness" and International Cooperation

In the introductory chapter of this volume, Solingen examines the impact of external factors on relations between the scientific community and the state. She suggests that a high degree of external conflict "leads to higher levels of investment in military-related scientific research, higher secrecy (including domestic), with respect to scientific research, and to lower levels of openness to international scientific interdependence."

In Israel, the military R&D system is indeed highly secretive, particularly with respect to such areas as nuclear research, missiles, and related technology. In contrast, however, basic science and civil R&D are very open. Israeli scientists publish extensively in international journals (among the highest rates per capita in the world). Basic research in areas that might seem sensitive because of their relationship to military sectors, such as laser isotope enrichment research and electro-optics, are not generally subject to formal government censorship or other restrictions. In many cases, however, the scientific community exercises a high degree of self-censorship in "sensitive" areas.

Although the Israeli economy is far more integrated and dependent on the world economy than that of the United States, the formal restrictions on the transfer of civil technology and knowledge that have developed in the United States have no counterpart in Israel. Indeed, some critics have warned against the relative freedom with which the Israel scientific community has "given away" new irrigation technology and agricultural developments, as well as innovations in the area of military and medical technologies. They argue that this process contributes to increased international competition that will lead to—or, in some cases, already has led to—reduced export earnings.

The percentage of joint research projects and publications with colleagues in other countries is very high: 27 percent of the papers written by Israeli scientists include international coauthors, compared with 10 percent for the United States, and a range of 15 to 20 percent for most Western European states.[23]

Government institutions and policies encourage scientists to travel and to participate in international exchanges and cooperative projects. At any given time, hundreds of Israeli scientists are on sabbatical leave in the United States alone.[24] With a population of five million and a GNP of $52 billion in 1991, Israel is by all measures a small country with very limited absolute resources for science and technology. Israel cannot conceivably produce its own atomic particle accelerators or space-based instruments to provide access to the "frontiers of science" in these areas. Similarly, large-scale funding for biomedical research and laboratories on the scale found in the United States or Western Europe is beyond the range of Israel's budgetary realities. Thus, in order to participate in the major R&D activities of the modern era, Israeli scientists must join in cooperative ventures.[25]

Most of Israel's scientific and technological links are concentrated in Western Europe and North America. A very large portion of scientific exchanges involve individuals and institutions from these regions. Formal government-to-government exchange programs and frameworks for cooperative research have been signed with United States, France, Holland, and other countries.

A significant percentage of R&D funds in Israel is provided in cooperative frameworks such as the U.S.-Israel Binational Science Foundation. Between 1958 and 1972, the United States funded foreign educational and research projects under the Public Law 480 program.[26] Since then, an endowment of $100 million was created for the Binational Science Foundation (BSF) with additional funds in similar frameworks for agricultural and industrial research (BIRD-F and BARD-F). Support is also obtained from direct funding through contracts from U.S. agencies such as the National Institutes of Health, Agency for International Development, National Aeronautics and Space Administration, Department of Education, and Department of Defense. As of 1985, over one-third ($26 million out of the $70 million) of the funds for basic research conducted in Israeli universities came from foreign sources.[27]

---

23. Shlomo Hershkowitz, ed., *Scientific Research in Israel: 1988,* (Jerusalem: National Research and Development Council, 1989).

24. Keynan, *Science and Israel's Future* (see n. 2), 47.

25. M. Carmi, *Research and Development Systems in Small Countries with Special Reference to Israel* (Jerusalem: National Council for Research and Development, 1979).

26. Under United States Public Law 480, the income derived from the sale of U.S. surplus food supplies to nonindustrialized countries was made available to the recipient country to support research.

27. Keynan, *Science and Israel's Future* (see n. 2), 49–50; see also Foundation for Basic

This basic condition is recognized by the government, which facilitates such cooperative activities through special travel funds and other arrangements and incentives. The NCRD publicizes and coordinates scientific exchanges and related agreements with other states and international bodies.

The dependence of the Israeli economy on imports and exports, and the technological-based export strategy have further highlighted the importance of international technological links.[28] As noted, over the past two decades, the percentage of technology-intensive exports, including defense and military products, has grown rapidly and the Israeli economy is now highly dependent on this sector. Exports center on computer industries, aeronautical equipment, electronics, pharmaceuticals, and communications systems. In many cases, Israeli firms have become subcontractors in major weapons systems developed by U.S. prime contractors. Given this structure, and the dependence on exports, the R&D community must necessarily be in close contact with its counterparts in the industrialized world.

The Effects of Regional Isolation

The emphasis on international links in the area of R&D is further heightened by the political and economic isolation of Israel in the Middle East. For most of its existence Israel has not had diplomatic or economic links with any of the states in the region, although under the 1979 peace treaty, relations were established with Egypt. The economic boycott of Israel imposed by most of the Arab world includes a secondary boycott, which limits trade with many industrialized countries, and Japan in particular. Between 1967 and 1990, Israel was politically and economically isolated from Eastern Europe, the Soviet Union, much of Africa, and parts of Asia.

However, as a result of the international prominence of Israeli scientists and their technological achievements, members of this community were generally far less isolated than other population groups. By virtue of their professional expertise and international contacts, scientists were often invited to participate in international conferences and exchanges in countries from which other Israelis were generally excluded, such as China, the Soviet Union, Eastern Europe, and even a few of the Arab countries that have continued to maintain a state of war with Israel.

Science and technology, or more precisely, scientists and members of the technological community, became a channel for diplomacy and international

---

Research, *Annual Report,* no. 12, 1986–87; and no. 14, 1988–89 (Jerusalem: Israel National Academy of Sciences, 1987–89); Israel National Academy of Science, *Activities of Scientific Research in Israel* (Jerusalem: Israel National Academy of Science, 1986).

28. For a theoretical discussion of this point, see Solingen in this volume.

acceptance for Israel. Scientists have often represented the state unofficially in international organizations and settings, and among scientists from states that do not have diplomatic relations with Israel. Professional meetings and conferences, as well as the activities of groups such as Pugwash, provided the setting for meetings between Israeli and Egyptian scientists before the 1979 peace treaty. Professor Michael Sela, from the Weizmann Institute, was among the first Israelis to be invited to the Soviet Union during the Gorbachev era. At the time, when few other channels existed, these scientists provided a means of communication for the government in meetings with political officials, including Gorbachev, and often carried diplomatic messages to and from Israeli political leaders. Similarly, the establishment of diplomatic and economic links with China and India began with scientific and technological exchanges and collaboration (in both the civil and military sectors).

To counter this political isolation, Israel is strongly represented in international scientific organizations, such as UNESCO, the scientific divisions of the International Atomic Energy Agency (IAEA), the European Molecular Biology Organization (EMBO, which Israeli scientists helped found), and others. Membership in such international organizations is seen as providing a form of international recognition when direct state-to-state relations have been absent.

Many scientists and political leaders see scientific and technological cooperation with neighboring Arab states, such as Egypt, Jordan, Syria, and Saudi Arabia as a basis for establishing cooperation and reducing the scope of the military and political conflict. Jordanian farmers have adapted many of the Israeli techniques for water conservation and drip irrigation, and this indirect contact is seen by Israelis as part of a broader "functionalist" approach to conflict reduction. Similarly, Neeman promoted the Mediterranean–Dead Sea Canal project as a means of establishing technical cooperation with Jordan. In the environment of the "cold peace" between Egypt and Israel, scientific exchanges and even some cooperation provide a means of increasing contacts. These exchanges include joint research projects in public health, marine sciences, agriculture, and physics.[29]

While such open and frequent international exchanges might be a source of tension or even "subversion" in closed political systems, this is not the case for Israel. Given the open and highly pluralist political system, there is little that foreign travel and exposure to external concepts or to the "free marketplace of ideas" can provide that is not already found in Israel.

Far from being a source of subversion, exchanges are seen as source of

29. In a publication appearing in *Physical Review Letters*, two Egyptian physicists acknowledged the assistance of a member of the Weizmann Institute faculty (see M. El-Nadi and O. E. Badawy, *Physical Review Letters* 61, no. 11 (12 September 1988): 1271–72).

political support for Israel. Visiting foreign scientists exposed to the realities of Israel, and those who meet Israeli scientists abroad, are expected to develop an understanding of and sympathy for their Israeli colleagues. Thus, in the Israeli case, there is no contradiction between a high degree of international interdependence and the political impact of globalization. On the contrary, these are seen as synergistic and mutually reinforcing.

Happy Convergence

In chapter 1 of this volume, Solingen uses the term *happy convergence* to define a situation characterized by a high degree of consensus between state structures and the scientific community. Happy convergence allows scientists to operate with "broad measures of internal freedom of inquiry" and provides this community "with relatively comfortable material rewards." Efforts by the state to intervene with and control the direction of science are resisted.

In the Israeli case, the evidence conforms closely with this model. The mutual dependence between science and the state has led to consensus and a high level of mutual cooperation. As discussed previously, leaders of the scientific community form part of the ruling elite, and participate actively in the decision-making processes in both the military and economic spheres. On a relative scale, the "material rewards" for scientists are high. Through various links, both informal and formal, these scientists determine policies, allocate funds, and set priorities on behalf of the state, while intervention by the state is very limited.

The close links between the scientific community and the military, and the unofficial diplomatic role of individual scientists in the face of Israel's formal isolation, tend to reinforce the situation of happy convergence between the state and the scientific community.

This is a community of "insiders", using R. Gilpin's terminology; the "outsiders" are few in number and have very limited influence.[30] With very few exceptions,[31] the Israeli scientific community forms a largely homogenous group, and it identifies strongly with the state and its overall policies. Although the community is divided, much as the broader polity is split, between "hawks" and "doves" with respect to policy options in the context of the Arab-Israeli conflict, scientists generally share the overall Israeli con-

30. R. Gilpin, *American Scientists and Nuclear Weapons Policy* (Princeton, NJ: Princeton University Press, 1962).

31. A few individual scientists, such as Israel Shahak, are active opponents of government policies. They lack any form of organizational support, and their political influence in the scientific community is essentially nil. Their political activities are generally outside the broad framework of Israeli politics and protest movements.

sensus regarding the centrality of Jewish sovereignty and national survival.[32] In international organizations such as UNESCO, the IAEA, and Pugwash, Israeli scientists work closely with the government to oppose Arab-sponsored anti-Israel initiatives.

At the same time, there have been two significant exceptions to homogeneity and the convergence of views on issues of national security and technology. The first concerned the case of atomic weapons development. The Israeli nuclear research program was developed in the 1950s under the joint auspices of the Ministry of Defense and the Israel Atomic Energy Commission (IAEC). In 1957, as the military potential of the program grew (in response, in large part, to the Egyptian efforts to develop missiles and chemical weapons), all of the members of the IAEC resigned with the exception of Professor David Bergmann, who was known to have close links with the military.[33] A few years later, as the military nature of this undertaking became more pronounced, a number of scientists created the Israel Committee for a Nuclear Free Zone. This group began a vigorous public discussion, but, in 1963, when the development of the Dimona facility seemed to slow, and Prime Minister Eshkol announced that Israel had no plans to produce nuclear weapons, this political activity subsided.[34]

In 1966, the structure of the IAEC was altered and Prime Minister Eshkol replaced Professor David Bergmann as chairman. A number of conflicting explanations have been presented for this move. According to Alan Dowty, the reorganization was the result of conflict within the IAEC and the government regarding the development of a nuclear deterrent. Dowty argues that Prime Minister Eshkol had accepted U.S. pressure to "freeze" development of the Dimona facilities, prompting (or perhaps forcing) Bergmann to resign.[35] At the same time, a group calling itself the Committee for the Nuclear Disarmament of the Arab-Israeli Region emerged. Fifty-two members of this group, which included scientists from Israel's major research institutions such as the Weizmann Institute and the Technion (including two

---

32. It should be noted that overall, the natural scientists (physics, biology, chemistry) tend to be more "hawkish" with respect to the Arab-Israeli conflict and the debate over policy options, while those from the social science and humanities tend to be more identified with the "Left" and groups such as Peace Now. During the 1988 general elections, academic signatories of Peace Now advertisements in newspapers were overwhelmingly from the social science and humanities faculties, while natural scientists formed the majority of signatories in advertisements for the right-wing Likud and Tehiya parties.

33. Alan Dowty, "Israel and Nuclear Weapons," *Midstream* 26 (November 1976): 10.

34. Ephraim Inbar, "Israel and Nuclear Weapons since October 1973," in *Security or Armageddon: Israel's Nuclear Strategy,* ed. Louis Rene Beres (Lexington, MA: Lexington Books, 1988), 62.

35. Alan Dowty, "Nuclear Proliferation: The Israeli Case," *International Studies Quarterly* 22, no. 1 (March 1978): 79–120.

former members of the IAEC who had resigned in 1957), published a petition calling on "the Government of Israel to take the international political initiative in order to prevent the proliferation of nuclear weapons in the Middle Eastern land mass."[36]

The level of conflict and protest within the scientific community regarding the development of a nuclear weapons option was limited and apparently ineffective. Since the period preceding the 1967 war, in which the conventional force of the combined Arab armies threatened the survival of the state, little criticism of the maintenance of a nuclear option has emerged in the country in general, and in the scientific community in particular. The case of Mordechai Vannunu, a former nuclear technician who provided detailed descriptions of the Dimona facility and operations to the *Sunday Times of London,* and was subsequently convicted for treason, failed to generate support among scientists, or any form of "collective action" as a professional community.

A second case of public dissent took place in 1992 around the issue of the Israeli military space program, and the development of reconnaissance satellites in particular. This effort was accelerated following the 1991 Gulf War, in which the threat of Iraqi missiles and chemical weapons paralyzed Israel for six weeks. After the war, the Israeli military recognized the need for an independent source of real-time intelligence on military deployments in the Arab world, and this led to a dedicated satellite reconnaissance program.

The former head of the Israeli Space Agency, Professor Dror Sadeh of Tel Aviv University, criticized this program, arguing that this technology was too expensive and not cost effective, given limited Israeli resources. He argued that this effort would divert scarce resources from the small-scale scientific R&D program, and could reduce cooperation in this area with other states, including the United States.[37] However, Sadeh has been largely isolated in this debate, and most members of the scientific community seem to actively support and participate in the government policy. This criticism generated even less support or public discussion than the earlier protests over nuclear policy had, but both cases demonstrated the openness of the Israeli society, and the absence of formal legal or political limitations on policy debate.

---

36. Committee for Nuclear Disarmament of the Arab-Israeli Region, "Keep Nuclear Weapons Out of Our Region," *New Outlook* (July-August 1966) 64–65. (*New Outlook* is a small publication that serves as a protest vehicle against government policies.) Additional articles on the question of nuclear weapons were published in the mid-1960s. See E. Livneh, "Israel Must Come Out for Denuclearization," *New Outlook* (June 1966): 44–47; Y. Nimrod and A. Korczyn, "Suggested Patterns for Israeli-Egyptian Agreement to Avoid Nuclear Proliferation," *New Outlook* (March 1966): 9–20; and V., Y. "Atoms and Middle East Tashkent," *New Outlook* (March 1966): 3–6.

37. *Ha'aretz* (Israeli daily), 21 May 1992.

Technological Enthusiasm

Given the high degree of interaction and interdependence, there is perhaps surprisingly little conflict within the scientific community, or between the state and the scientific community. In the military sphere, while there may be some disagreement about the means and appropriateness of some projects, active protests and opposition to policies is largely nonexistent (with the exception of the nuclear case discussed above.) In Israel, there is no equivalent to groups such as the Union of Concerned Scientists (UCS) or the Federation of American Scientists (FAS). There are also no well-known individuals who serve as "opposition" scientists, and scientific meetings generally have no political content or debate.

The close relationship and interdependence between the state and the scientific community has increased the tendency in Israel for policy decisions based on "technological enthusiasm," as exhibited in the cases of the Lavi combat aircraft project and the Mediterranean–Dead Sea Canal project. The Israeli political leadership, as noted above, has a strong tendency to view science and technology without differentiation as a source of economic growth, social development, national security, and national prestige. "Technological fixes" are often proposed for difficult social problems, and projects that are in some way linked to R&D tend to be endorsed enthusiastically without detailed cost-benefit assessments. Given the uncertainties involved, instead of assessments, such projects are rationalized under the assumption that all spending in R&D ultimately contributes to the national good in some way. The concept of spillover is frequently used to rationalize decisions that do not seem to be cost effective. The allocation policies of the OCS and Israel Investment Center are based on a similar approach.

The Israeli scientific and technological community has generally reinforced this undifferentiated "technological enthusiasm." In part, this is a reflection of the shared values of this community; indeed, most are "technological enthusiasts" themselves, so it is not surprising that opposition to such policies is rare. In addition, the scientists and engineers often depend directly on these policies and resulting allocations; since many members of this community receive subsidies from the OCS, IIC, or other source, they are unlikely to criticize related policies.

While state-scientists relations may have been originally based on the sociological links between the political leadership and the individual scientists, the current relationship is clearly influenced, or perhaps dominated by political-economic issues.[38] Both factors play a role in explaining the failure

38. See Sanford A. Lakoff, *Knowledge and Power: Essays on Science and Government* (New York: Free Press, 1966); Don K. Price, *The Scientific Estate* (Cambridge, MA: Harvard

of the scientists to participate visibly in the public debates on the Lavi, the Arrow ATBM missile, energy sources, and the environment. To some degree, the absence of national institutions or channels may be a contributing factor. In general, however, the shared values and perspectives would seem to provide the fundamental explanation for the absence of structural or ideological conflict.

A Comparative Perspective

The relationship between scientists and state, like many other aspects of Israeli society and political economy, is somewhat anomalous, if not unique. The Western European cultural and social structure, and the emphasis on the independence of the scientific endeavor, exists within a political-economic structure that is characteristic of the developing world, in which mobilization of scientific resources is the norm.

In addition, the extreme importance of national security and the continuing concern for the survival of the state is highly unusual among industrialized and most industrializing states. (The closest analogy is perhaps with Pakistan, but even here, the military as well as the political threat to the survival of the nation is not as pronounced as the Israeli case.) The resources that are devoted to defense and national security (over one-third of the GNP) and the importance of science and technology in Israeli strategy are scarcely matched anywhere. The degree of Israel's international and regional isolation is also highly unusual.

Economically, the availability of scientific and technological resources and the role of these factors in industrialization, economic development, and export sales are also extremely pronounced. Both Brazil and India have also placed an emphasis on technology-oriented development and industrialization, but not nearly to the extent as these are emphasized in the Israeli case.

As a result, the role of the scientific and technological community in Israeli society is more pronounced than in most other cases. As noted with respect to Arens and Neeman, Israeli scientists and engineers not only have access to the decision-making elite; they are a central component of this elite. These figures play a major role in technology-related issues including the development of new advanced weapons systems and technology-intensive economic investments, and they are also involved in all aspects of policy-making in Israel. (As noted above, Neeman and Arens have held important positions in the government.)

In many other industrializing states, the scientific and technological en-

University Press, 1965); Dorothy Nelkin, "Technology and Public Policy" in Spiegel-Rosing and De Solla Price (see n. 6).

terprise is culturally alien. In contrast, in Jewish society in general, and Zionist ideology in particular, the scientific enterprise is widely valued and respected (which, as noted above, has contributed to the tendency toward "technological enthusiasm"). Even among the religious Israelis, who constitute some 20 percent of the society, there are few clashes between science and traditional values, and many physical scientists, engineers, computer experts, and physicians are members of the religious community.

The close links between scientists and the state and the general identity of interests and views has meant that there has been no conflict or perceived need for limitations or controls on scientists and their activities. The state generally sees its role as a resource provider for scientific and technological activities, rather than as a source of control and limits. In many cases, the close ties between the scientists and the military have allowed the former a great deal of freedom in new weapons projects and similar activities (as in the case of the Lavi, for example). State involvement in agenda setting in the civil sector is also highly circumscribed, except to the extent that the scientific backers of specific projects seek to mobilize the state for support (as in the case of the Mediterranean–Dead Sea Canal project, for example).

As a result of these close ties and common perceptions and interests, the Israeli case may perhaps be explained at least as well in terms of the sociological role of scientists in society as in terms of the political-economy models that are useful in most other cases.[39] While the role of technology in the economy has been and will continue to be of central importance, the changing nature of the political economy and the evolution of a more decentralized, pluralistic, and market-oriented system has not and is not likely to effect the close relationship between scientists and the state.

Perhaps most importantly in a comparative context, Israel's position in the region and the international system continues to be highly anomalous. While peace talks began in Madrid in October 1991, Israel remains largely isolated from most of its neighbors. The threat of war continues and has been reinforced by the Iraqi missile attacks during the Gulf War. The importance of technology in maintaining the military balance, and the contribution of this factor to the close convergence between the state and the scientific community is still prominent in the perceptions of both the political and scientific elite. As the technological arms race in the region accelerates, not only in the deployment of advanced conventional weapons, but also with the proliferation of ballistic missiles and chemical, biological, and nuclear warheads in Iran, Iraq, Syria, Libya, and Algeria, the close cooperation between state and science is likely to grow even more pronounced.

39. Joseph Ben-David, *The Scientist's Role in Society: A Comparative Study* (Englewood Cliffs, NJ: Prentice Hall, 1971).

Prospects for Change

In a broad sense, the relationship between science and the state in Israel can be explained by the particular factors that characterize the Israeli environment; the emphasis on the role of science and technology in Jewish and Zionist values and ideology, the central importance of technology to national security (which is by all measures, far greater than in other industrialized or industrializing states), the role of science and technology in the rapid transformation of the economy and society, and the role of science and technology in breaking Israel's diplomatic isolation in the international system.

As demonstrated previously, the interactions between the political processes and the scientific community are generally consistent with the models and theories of state-scientists relationships. The openness of the Israeli political system, and close interdependence between the state and the scientific community have contributed to a relationship based on happy convergence. The scientific community is part of the Israeli elite, and as part of that elite, plays an important role in determining economic and security policy. Scientists are involved in determining both policy for science and the role of science in policy.

As a result of the close links between scientists and the state, sharp conflicts have been avoided, and Israeli scientists are able to function internationally with minimal restrictions. Despite the importance of military technology and the role of scientists in the national security structure, the activities and international links of this community have generally not been subject to government-imposed limitations. On the contrary, the state has supported the international activities of scientists.

There are no indications that would point to changes in the elite role of scientists, or in the homogeneity and consensus approach to science and technology that have prevailed for over four decades. As long as the basic political and economic constraints that shape the structure of the Israeli political economy are constant, the interaction between scientists and the state are unlikely to change radically.

Contributors

FRANK R. BAUMGARTNER is associate professor of political science at Texas A&M University. His most recent publications include *Agendas and Instability in American Politics* (with Bryan Jones, 1993), and *Conflict and Rhetoric in French Policymaking* (1989), as well as a variety of articles on the French and American politics and policymaking in the *American Political Science Review, Journal of Politics, American Journal of Political Science, Legislative Studies Quarterly,* and *Comparative Politics.*

WENDY FRIEMAN is director of the Asia Technology Program at Science Applications International Corporation (SAIC) in McLean, Virginia, where she is responsible for research on science, technology and industry in East and Southeast Asia. In addition to directing the Japanese Technology Evaluation Program, Ms. Frieman has written extensively on Chinese research and development, Chinese military modernization, and Sino-Japanese technology transfer. Ms. Frieman is coauthor of the book *Gaining Ground: Japan's Strides in Science and Technology,* and is currently working on a book examining the science and technology boom, and its impact on the emerging economies of Asia. She has traveled and conducted field work in Europe, Japan, China, and Southeast Asia.

PAUL R. JOSEPHSON teaches in the Department of Science, Technology, and Society at Sarah Lawrence College, Bronxville, NY. He is the author of *Physics and Politics in Revolutionary Russia* (1991), and a forthcoming history of Akademgorodok, entitled *Oasis in Siberia: The Soviet City of Science.*

ASHOK KAPUR is professor of political science at the University of Waterloo, Ontario. Born in Lahore, he is a Canadian citizen. He received his B.A. (Honors) from Punjab University, India, his M.A. from the George Washington University, and his Ph.D. from Carleton University, Ottawa. He served as a member of the United Nations Committee to Study Israeli Nuclear Armament in 1980–81. His publications include *India's Nuclear Option: Atomic Diplomacy and Decision-Making* (1976), *International Nuclear Proliferation: Multilateral Diplomacy and Regional Aspects* (1979), *The Indian Ocean: Regional and International Power Politics* (1983), and *Pakistan's Nuclear Development* (1987), and *Diplomatic Ideas and Practices of Asian States* (1991), and *Pakistan in Crisis* (1991). *India and Her Neighbors* (with A. J. Wilson), is forthcoming. An article on South Asian perspectives on nuclear proliferation also appears in *Pacific Affairs* (Summer 1984).

MORRIS F. LOW teaches at Monash University in Melbourne and is currently deputy director of the Japanese Studies Center there. He has published widely in the history of Japanese science, especially physics. Recent articles include "The History of East Asian Science: State of the Art," *Studies in History and Philosophy of Science,* (forthcoming); "Japan's Secret War?: "Instant" Scientific Manpower and Japan's World War II Atomic Bomb Project," *Annals of Science* (1990); "The Butterfly and the Frigate: Social Studies of Science in Japan," *Social Studies of Science* (1989); and "Accounting for Science: The Impact of Social and Political Factors on Japanese Elementary Particle Physics," *Historia Scientiarum* (1989). He has also recently completed a major study of the role of physicists as policymakers in Japan.

FRANK R. PFETSCH studied economics and political science at the universities of Paris, Turin, Heidelberg, and at Harvard University. From 1964 to 1976 Dr. Pfetsch worked in research institutes and was a consultant to the West German Ministry for Scientific Affairs, and to UNESCO for various science policy administrations in South America, Africa, and Asia. Since 1976 Dr. Pfetsch has been a professor of political science at the University of Heidelberg and was a visiting professor at the University of Pittsburgh and at Kyung Hee University in Korea. Among his publications are the books *Zur Entwicklung der Wissenschaftspolitik in Deutschland* (1974), and *West Germany: Internal Structures and External Relations* (1988).

SIMON SCHWARTZMAN was born in Belo Horizonte, Brazil, where he did his first studies in sociology and political science. He teaches political science at the Universidade de São Paulo, where he is also the scientific director of the Research Group of Higher Education. He holds a Ph.D. from the University of California, Berkeley, and is the author of several books, including *Bases do Autoritarismo Brasileiro* (1988), and *Formação da Comunidade Científica no Brasil* (1979). An English version of the latter book, *A Space for Science: The Development of the Scientific Community in Brazil,* was published in 1991.

BRUCE L. R. SMITH has been a member of the senior staff of the Brookings Institution since 1980. He received his Ph.D. from Harvard University, and served as director of the Policy Assessment Staff, Bureau of Oceans, International Environmental and Scientific Affairs, in the U.S. Department of State. He was professor of public law and government at Columbia University from 1966 to 1979 and has also taught at Harvard, UCLA, MIT, Syracuse and Johns Hopkins Universities. His scholarly publications include *The RAND Corporation* (1966); *The Dilemma of Accountability in Modern Government* (with D. C. Hague, 1971); *The Politics of School Decentralization* (with G. R. LaNone, 1973); *The New Political Economy* (1975); *The State of Academic Science* (with J. J. Karlesky, 1977); *The Higher Civil Service in Europe and Canada: Lessons for the United States* (1984); and *The State of Graduate Education* (1985). His most recent books are *The Advisers: Scientists in the Policy Process* (1992); *The Next Step in Central America* (1991); and *American Science Policy since World War II* (1990).

ETEL SOLINGEN teaches international relations at the University of California, Irvine, and received her Ph.D. at the University of California, Los Angeles. Her research interests are in international relations theory and in the comparative political economy of science and technology. She is the author of several articles on foreign economic policy and industrial development and nuclear politics in Latin America and the Middle East, including contributions to *International Organization, International Studies Quarterly,* and *Comparative Politics.* She is now completing a manuscript entitled *Bargaining in Technology: Industrial Policy and the Nuclear Sector in Argentina and Brazil.*

GERALD M. STEINBERG is a senior lecturer in political science at Bar-Ilan University, Ramat Gan, Israel. He received his Ph.D. in government from Cornell University and specializes in the relationship between science and politics. He is the editor and primary author of *Lost in Space: The Domestic Politics of the Strategic Defense Initiative,* and his articles on large-scale national R&D projects and Israeli defense industries have appeared in *Policy Sciences, Comparative Politics,* and other journals.

DAVID WILSFORD is associate professor of international affairs at the Georgia Institute of Technology. He is the author of *Doctors and the State: The Politics of Health Care in France and the United States* (1991), as well as a number of articles on the policy-making processes of the advanced, industrial democracies.